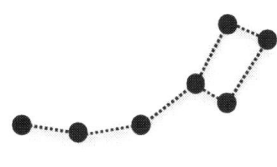
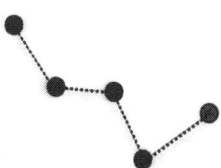

THE
CONSTELLATIONS
HANDBOOK

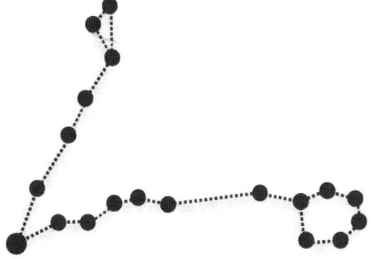
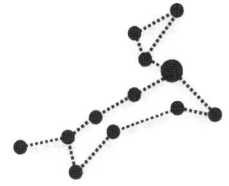

THE CONSTELLATIONS HANDBOOK

GALACTIC HUNTER

ANTOINE & DALIA GRELIN
GALACTIC-HUNTER.COM

CONTENTS

FOREWORD	6
ASTRONOMERS' BIOGRAPHIES	8

CLAUDIUS PTOLEMY 12

ANDROMEDA RELATED CONSTELLATIONS	14
WATER RELATED CONSTELLATIONS	26
ORION RELATED CONSTELLATIONS	36
CONSTELLATIONS IN PAIRS	50
HERCULES RELATED CONSTELLATIONS	56
URSA MAJOR RELATED CONSTELLATIONS	72
SUMMARY	78

JOHANNES HEVELIUS 82

ANIMALS	84
OBJECTS	89
SUMMARY	92

NICOLAS-LOUIS DE LACAILLE	94
AGE OF ENLIGHTENMENT TOOLS	96
ARGO NAVIS	106
SUMMARY	112
PETRUS PLANCIUS	114
PRE-VOYAGE CONSTELLATIONS	116
EAST INDIES CONSTELLATIONS	119
POST-VOYAGE CONSTELLATIONS	130
SUMMARY	132
INDEX / PERSONAL NOTES	134
END WORD	142
CREDITS	144

FOREWORD

Welcome to **THE CONSTELLATIONS HANDBOOK**. The purpose of this guide is to teach you ways to memorize the constellations easily.

We are Antoine and Dalia Grelin, a couple that have been enjoying the dark skies of the Nevada desert for years.
We enjoy teaching astronomy to as many people as possible, and as a result started a YouTube channel, Galactic Hunter, to make videos about photographing Deep Sky Objects.

Have you ever wished that you could look up to the sky, point to any pattern of stars, and name the constellation? We have tried several methods to learn those patterns but find it hard to remember. Let's face it, constellations are **difficult** to memorize!

During a beautiful, starry night, we met another astronomer named Francisco. We rarely meet anyone during our outings, but that night changed how we view constellations.

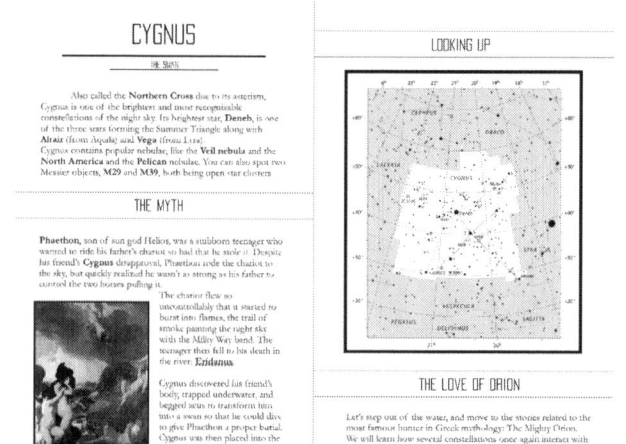

Francisco pointed out the constellation of **Corvus**, and explained to us the Greek mythology behind the crow we see in the sky. The story led to a neighboring constellation, **Crater**, and continued on to **Hydra**! Francisco, using memorable sets of words, succeeded in imprinting the story and location of three constellations we vaguely knew about!

A few days later, we became more and more interested in the stories behind the constellations, and realized that all of them (at least the ones introduced by the Greeks) had their own tale and interacted with one another.
It is by learning these stories that the names of the constellations, their shape, and their place in the heavens started making sense to us…

…And that is why this book was created!
In the following pages, you will find all **88** constellations. They are divided into four main groups, depending on which astronomer introduced them.

Each constellation page has a map, a description, and more! For the constellations that come from the Greek mythology, know that many of them are linked to several different stories. To avoid confusion and facilitate our goal of remembering them all, we chose to tell the stories that interact the most with one another and make the most sense with the constellation.

Let's introduce the four main astronomers of this book, and discover the 88 constellations we see today in our night sky!

Clear Skies,
Galactic Hunter

Ptolemy being guided by Urania

Claudius PTOLEMY was a Greco-Roman astronomer, mathematician, and geographer born in 100 AD. Ptolemy did not travel much, and lived his entire life in Alexandria, Egypt until his death in 168 AD.

Ptolemy had a huge impact on how we see our night sky today. He was the first to publish a full list of groups of stars forming a pattern, known as constellations.

His book, the Almagest, was largely written using **Aratus** and **Hipparchus**' work. The Almagest is the oldest, surviving constellation book, therefore, Ptolemy is credited for almost all the constellations seen in the northern hemisphere. Because Ptolemy did not like traveling, he was only able to list the stars he could see from his home in Alexandria.

Ptolemy originally introduced 48 constellations, but only 47 survived, as Nicolas-Louis de Lacaille, one of the four main astronomers we will talk about in this book, cut one of Ptolemy's constellations, Argo Navis, into three smaller ones.

The constellations in the Almagest were named after beasts, heroes, and objects from the Greek mythology. They can easily be remembered by learning the story behind them, and how they interact with one another.

Johannes HEVELIUS was a Polish astronomer born on January 28, 1611.

Hevelius spent his early years brewing beer, but became greatly interested in astronomy and even built his own observatory named "Star Castle", which was often visited by Polish King **John III Sobieski**.

Edmond Halley visited him in 1679, and was sent by none other than **John Flamsteed** and **Robert Hooke** to persuade Hevelius to use telescopes for his measurements. Hevelius hated telescopes, and told Halley that his sextant and alidade were enough for all his work. Sadly, the same year Halley visited him, Hevelius' Star Castle became engulfed in flames.

The incident had a negative impact on Hevelius' health, and he succumbed on January 28, 1687, the day of his 76th birthday.

Johannes Hevelius is renowned as the last astronomer to have a major impact on the world without using a telescope. He is also credited as the founder of lunar topography, and of course is known as one of the fathers of the constellations.

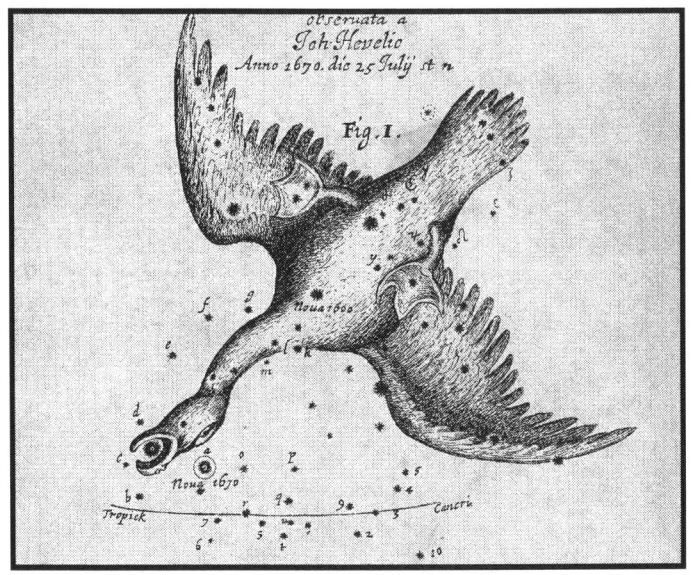

Location of the 1670 nova in Cygnus, by Johannes Hevelius.

Born in March 1713, French astronomer **Nicolas-Louis de LACAILLE** started his life as an abbot. He then became a professor of mathematics at the University of Paris.

Some of Lacaille's students included Antoine Lavoisier and Jean Sylvain Bailly, who both died via guillotine during the French Revolution.

In 1750, The French Academy of Sciences commissioned Nicolas-Louis de Lacaille to record observations of the southern sky. He set off, and on April 19th, 1751, the astronomer arrived at the Cape of Good Hope, in South Africa.

Using a quadrant, a sextant, and a tiny, 26-inch focal length telescope, Lacaille observed the sky every night for over a year. He recorded the position of more than 10,000 stars and 42 deep sky objects, some of which appear in the Messier catalog.

Although the astronomer was accompanied by his technician, **Mr. Poitevin**, only his dog **Gris-Gris** would be by his side during his night time recordings.

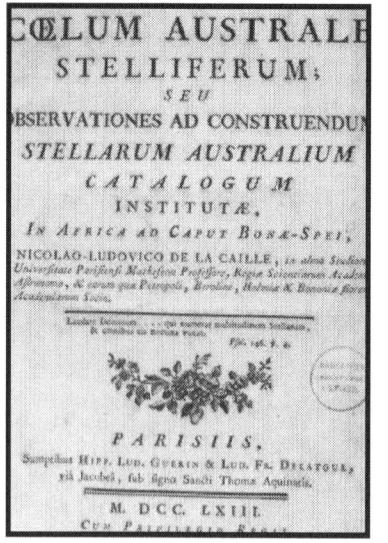
Coelum Australe Stelliferum (1763)

Lacaille went back to France in June 1754, and immediately began to organize all his notes to write his star catalog, alone. He introduced 14 new constellations, and cut one of Ptolemy's in three, meaning Lacaille has a total of 17 constellations recognized today. Sadly, the astronomer died at the age of 49 by overwork. His star catalog, the *Coelum Australe Stelliferum*, was published posthumously in 1763, one year after his death.

Uranometria, by Johann Bayer

In 1595, Dutch globe maker **Petrus PLANCIUS** contacted **Pieter Keyser**, a boat navigator, to help catalog the position of the southern stars while traveling to the Indies. At the time, no detailed map of the southern night sky existed. This meant navigation by sea would be more difficult than it would be in the northern hemisphere. Keyser, helped by another Dutch explorer, **Frederik de Houtman**, recorded approximately 130 stars.

Out of the 248 sailors embarking on the journey with them, only 81 survived and came home in 1596. Sadly, Pieter Keyser perished before the crossing. His colleague, de Houtman, was amongst the survivors and delivered the observation notes to Plancius, who was able to create 12 new constellations in the southern skies. Plancius most likely used Keyser's notes to come up with the names for each, most of which were based on animals that explorers encountered in the Indies (Water Snake, Peacock, Bird of Paradise…)

Seven years later, German cartographer Johann Bayer created his famous star atlas, the Uranometria, which included, for the first time ever, the 12 new constellations made by the Dutch, as well as the 48 Ptolemy introduced.

In 1612, Plancius came up with eight extra constellations, however, only two were accepted by other astronomers (Monoceros and Camelopardalis).

In total, 16 constellations were made popular by the Dutch explorers.

CLAUDIUS
PTO

Uranometria, by Johann Bayer

In 1595, Dutch globe maker **Petrus PLANCIUS** contacted **Pieter Keyser**, a boat navigator, to help catalog the position of the southern stars while traveling to the Indies. At the time, no detailed map of the southern night sky existed. This meant navigation by sea would be more difficult than it would be in the northern hemisphere. Keyser, helped by another Dutch explorer, **Frederik de Houtman**, recorded approximately 130 stars.

Out of the 248 sailors embarking on the journey with them, only 81 survived and came home in 1596. Sadly, Pieter Keyser perished before the crossing. His colleague, de Houtman, was amongst the survivors and delivered the observation notes to Plancius, who was able to create 12 new constellations in the southern skies. Plancius most likely used Keyser's notes to come up with the names for each, most of which were based on animals that explorers encountered in the Indies (Water Snake, Peacock, Bird of Paradise...)

Seven years later, German cartographer Johann Bayer created his famous star atlas, the Uranometria, which included, for the first time ever, the 12 new constellations made by the Dutch, as well as the 48 Ptolemy introduced.

In 1612, Plancius came up with eight extra constellations, however, only two were accepted by other astronomers (Monoceros and Camelopardalis).

In total, 16 constellations were made popular by the Dutch explorers.

CLAUDIUS
PTO

2nd Century
Northern Hemisphere
Greek Mythology

LEMY

48
CONSTELLATIONS

ANDROMEDA

THE CHAINED LADY

Spanning over 1,400 full moons and located just north of the celestial equator, the Andromeda constellation will be our starting point in this guide.
Andromeda was first described by the Greco-Roman astronomer **Ptolemy**, along with several other constellations drawn in the 2nd century.

The constellation of Andromeda is home to the largest, brightest, and closest, spiral galaxy visible in the night sky with the unaided eye: **Messier 31** (or the "Andromeda Galaxy").

Two other Messier objects can be found within the constellation: **M32** and **M110**, which can be seen in the same field of view as M31 through most instruments.

THE ANDROMEDA GALAXY

The constellation of Andromeda would not be as interesting if it was not in the presence of the Andromeda galaxy. Unsure of what to do tonight? Grab a pair of binoculars and drive out to a dark location, as far away as possible from city lights. Although it can be tricky to spot at first, this magnificent galaxy is easily found by star-hopping.

Point your binoculars to Alpheratz, which is the brightest star in the constellation and it is also one of the stars forming the great square of **Pegasus**! Spot the next bright star (to the left of Alpheratz on the map), and continue left to land on Mirach. All you have to do now is turn at a 90 degree angle to the right, to have one of the best sights of your life.

LOOKING UP

- Best observed in **Fall**
- The **Andromedids** meteor shower peaks every November
- The brightest star in the constellation is **Alpheratz**

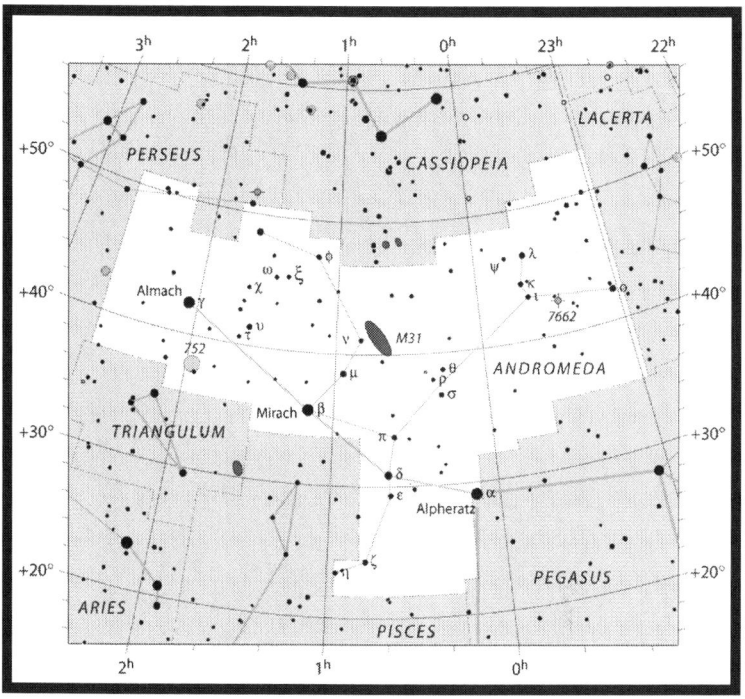

NOTABLE OBJECTS	M31	M32	M110	NGC 891	NGC 7662
TYPE	GALAXY	GALAXY	GALAXY	GALAXY	NEBULA
CATEGORY	SPIRAL	ELLIPTICAL	ELLIPTICAL	BARRED SPIRAL	PLANETARY

THE RESCUE OF ANDROMEDA

Andromeda was a princess, daughter of Cepheus, the King of Aethiopia, and Queen Cassiopeia.

Cassiopeia was a proud, but arrogant mother. She put herself in danger when she shouted that both she and her daughter were more beautiful than the sea nymphs, the Nereids.

This infuriated the god of the sea, Poseidon, who sent the sea monster **Cetus** to ravage the land of Aethiopa. Not knowing what to do to stop this attack on their people, the king and the queen traveled to the Libyan desert to find an oracle and ask for advice. The answer they received was heart-breaking.
The only way to stop Cetus was to surrender Andromeda and let the sea monster eat her. Anguished, the king and queen were convinced by the people of Aethiopia to concede, and **Cepheus** chained his daughter to a rock near the water and left.

As the sea monster approached the powerless princess, the hero **Perseus** and his flying horse **Pegasus** appeared over the scene. They were flying back after having decapitated the Gorgon, Medusa, when Perseus spotted the beautiful woman and immediately fell in love with her. The duo landed and, with a promise of marrying Andromeda, used Medusa's head to turn Cetus into stone thus slaying the monster.

Cepheus, Cassiopeia, and the citizens of the land of Aethiopia thanked Perseus for his courage. He later married princess Andromeda with her father's full support.

Poseidon, enraged at the outcome of his plan, decided to chain Cassiopeia to the sky, trapped for life. The queen spends half of each year upside down in the heavens as punishment for her words.

CEPHEUS

KING OF AETHIOPIA

Home to a diffuse nebula containing the largest discovered protostar, NGC 7538, Cepheus is a constellation next to Andromeda and Cassiopeia. Its brightest star is **Alderamin**, with a magnitude of 2.51.

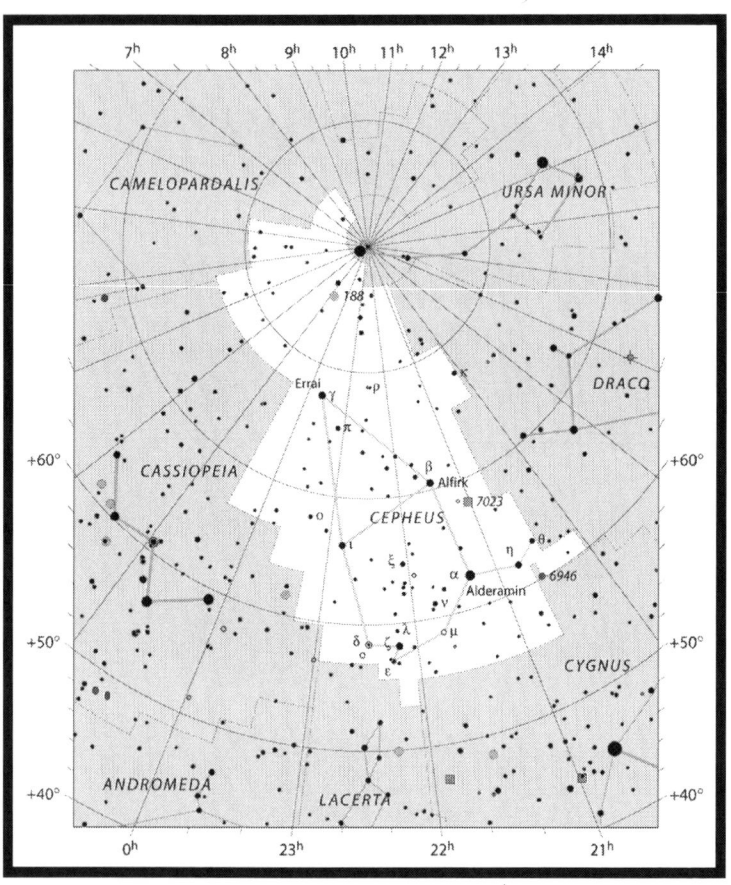

- Best observed in **Fall**
- There are no meteor showers in Cepheus
- The brightest star in the constellation is **Alderamin**

CASSIOPEIA

QUEEN OF AETHIOPIA

Cassiopeia is one of the most recognizable constellations in the night sky, and it is also one of the brightest. Cassiopeia can be found from its asterism: the 5 bright stars that form the shape of a "W" or "M".

The constellation contains two Messier objects (open clusters **M52** and **M103**) as well as several deep sky objects, including the famous **Heart and Soul nebulae**.

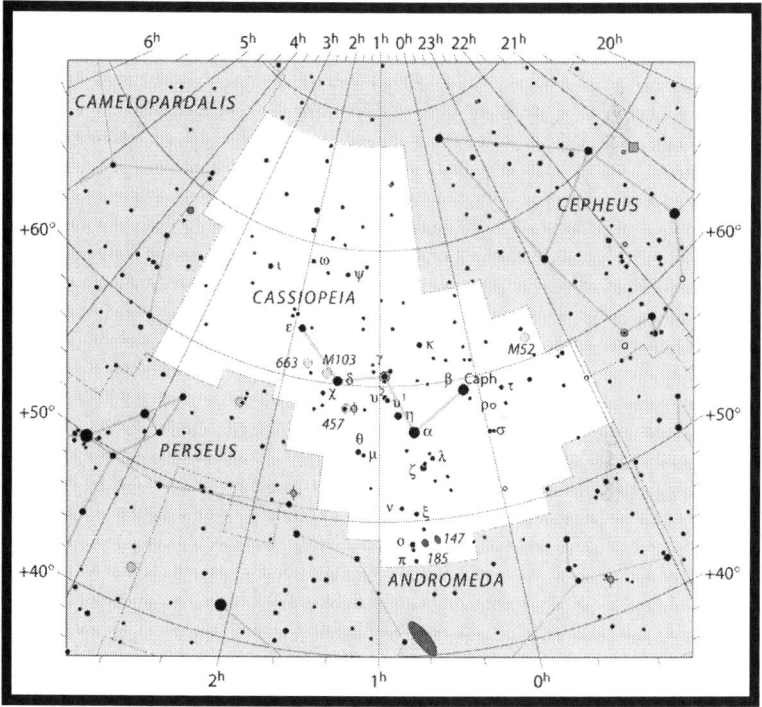

- Best observed in **Fall**
- The **December Phi Cassiopeiids** meteor shower peaks in early December
- The brightest star in the constellation is **Schedar**

PERSEUS

THE MONSTER SLAYER

Perseus is another of Ptolemy's constellations and is located near Andromeda and Cassiopeia. Being in the Milky Way's galactic plane, the constellation of Perseus is home to several deep sky objects, the **Little Dumbbell** nebula (M76) and the **Double Cluster** in Perseus (NGC 869 and NGC 884).

Comet Swift-Tuttle is responsible for the impressive Perseids meteor shower.

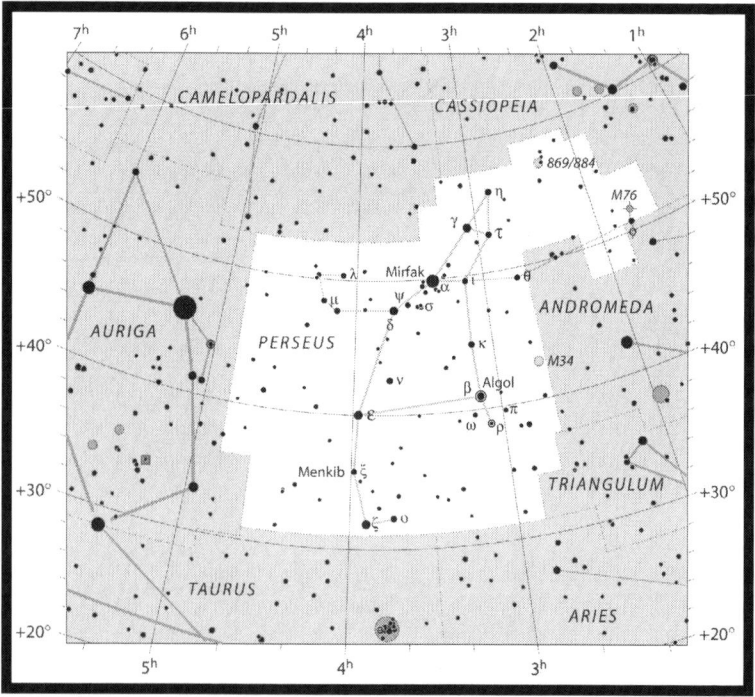

- Best observed in **Winter**
- The **Perseids** meteor shower peaks in mid-August
- The brightest star in the constellation is **Mirfak**

PEGASUS

THE FLYING HORSE

Pegasus is the seventh largest constellation in the entire night sky. It is easily recognizable by its enormous, near-perfect square of stars that rises very high in the night.
Pegasus' "square" is one of the two best places to start when star-hopping to find the **Andromeda galaxy**, the other being Cassiopeia's "W" asterism.
The constellation of the winged horse contains one Messier object, globular cluster **M15**, and several other deep sky objects such as the famous group of galaxies, **Stephan's Quintet**.

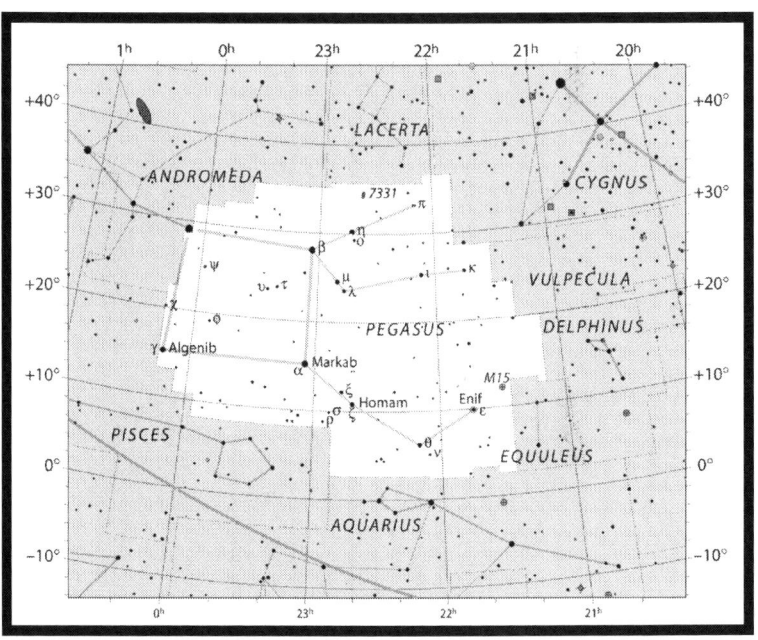

- Best observed in **Fall**
- The **Eta Pegasids** meteor shower peaks May 30th
- The brightest star in the constellation is **Enif**

CETUS

THE SEA MONSTER

The last constellation linked to Andromeda's story.

Cetus lies less than a quarter of a degree away from the ecliptic line, which means that the sun, the moon, and the planets are sometimes part of the sea monster for a short time.

It is very faint, and it is home to several deep sky objects. The only popular one amongst amateur astronomers and astrophotographers is the barred spiral galaxy **Messier 77**.

M77, also known as "Cetus A", was discovered in 1780 by French astronomer **Pierre Méchain**. It is located almost 50 million light years away and has a diameter of about 150,000 light years.
Cetus A is the central member of the M77 Group, which contains several galaxies bound by gravity.

MIRA

See the star labeled "Mira" on the map, just under M77?
It is a pulsating variable star by the name of Omicron Ceti, best known as "Mira". Its name is Latin for "Wonderful", and was named by **Johannes Hevelius**, as taken from his 1662 notes, *Historiola Mirae Stellae.*

The brightness of the star varies immensely in a 332-day period, ranging from magnitude **2.0** to **10.1**, thus often being completely invisible to the naked eye.
This can make the constellation tricky to find for those unfamiliar with Cetus, as one of its main stars disappears from time to time.

LOOKING UP

- Best observed in **Fall**
- There are **three** meteor showers associated with Cetus
- The brightest star in the constellation is **Deneb Kaitos**

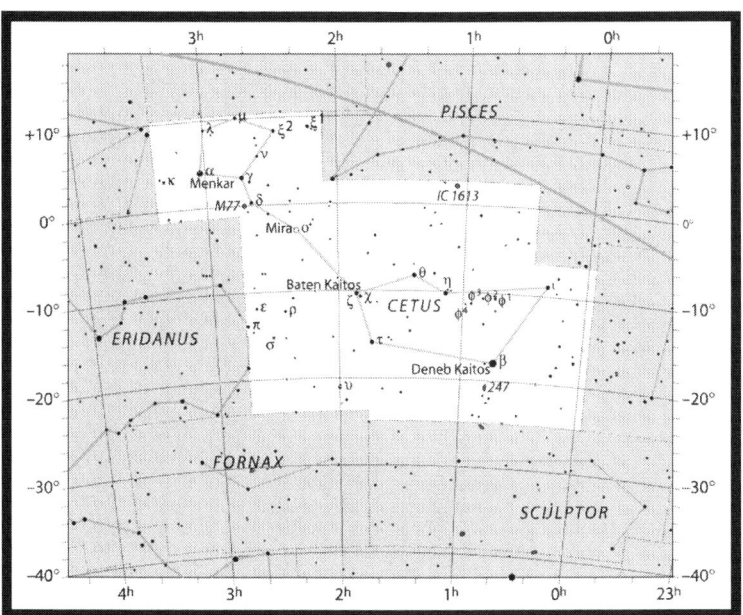

STEPPING INTO WATER

This concludes Andromeda's section and the five constellations related to the princess' story.
Now to leave the sea monster and dive into the myths of other water related constellations. Not to worry, they are nothing like Cetus!

WATER ESCAPADES

After the Olympian gods won the war against the Titans and the Giants, Mother Earth Gaia and Tartarus, trapped in the Underworld like the rest of their kind by Zeus, procreated a monster: Typhon.
Typhon was huge, cruel, and lawless. He had hundreds of heads with flaming eyes on each one, and was the most frightening beast to roam the Earth at the time.
Gaia, full of anger and seeking revenge for the lost war, sent Typhon to slay the gods.

Resting near a water source, the goat-headed god of the wild, Pan, was the first to hear Typhon approaching. Without a second to waste, Pan alerted the other gods then jumped into the river. Unable to swim quick enough, Pan transformed his lower body into a fish tail and successfully escaped the monster. Pan can be seen in the night sky as the constellation **Capricornus**.

Following the river, Typhon was getting closer and closer to the goddess of love, Aphrodite, and her son, Eros. Both managed to escape their fate by jumping into the water and transforming themselves into fish. Aphrodite, in order to keep her son close in the panic, tied herself to him with a rope and fled. They can be seen as the constellation of **Pisces**.

Typhon was later defeated by Zeus using thunderbolts brought to him by his eagle. It is said that the thunderbolt used to finish Typhon is now in the sky, as the constellation of **Sagitta**. The eagle (**Aquila**) can be seen guarding the thunderbolt.

Not far from Aquila is **Aquarius**. Aquarius comes from the son of King Tros, the young and beautiful boy, Ganymede. Because of his beauty, Zeus sent his eagle to abduct the boy (see left) and bring him back to Olympus so he would serve as water-bearer to the gods. Ganymede can be seen in the night sky as pouring water into **Piscis Austrinus**, the Great Fish constellation.

CAPRICORNUS

THE GOAT

To continue with the constellations introduced by the Greco-Roman astronomer, Ptolemy, the first in this series of constellations is Capricornus. Latin for "The Horned Goat", it represents the Greek god of the wild, Pan, who had goat legs and horns, then grew a fish tail to save his life.
Capricornus contains several galaxies and clusters. Famous deep sky objects located in its boundaries are **Messier 30**, a beautiful globular cluster, which lies just one degree south of the galaxy group **NGC 7103**.

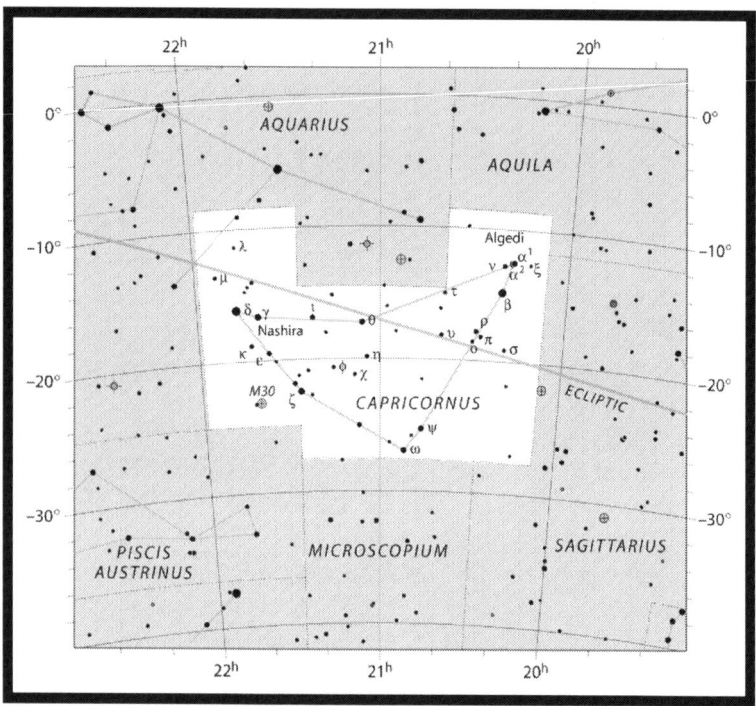

- Best observed in **Fall**
- There are **five** meteor showers associated with Capricornus
- The brightest star in the constellation is **Delta Capricorni**

PISCES

THE TWO FISH

Located between Aquarius, Cetus, Aries, and Pegasus is Pisces. It is a large constellation known since the 2nd century. The name "Pisces" is Latin for the plural form of "Fish", as it represents **Eros** and his mother **Aphrodite**, who, similarly to Capricornus, had to morph into fish to escape the monster Typhon.

Messier 74, a spiral galaxy discovered in 1780, can be spotted in the constellation of the two fish.

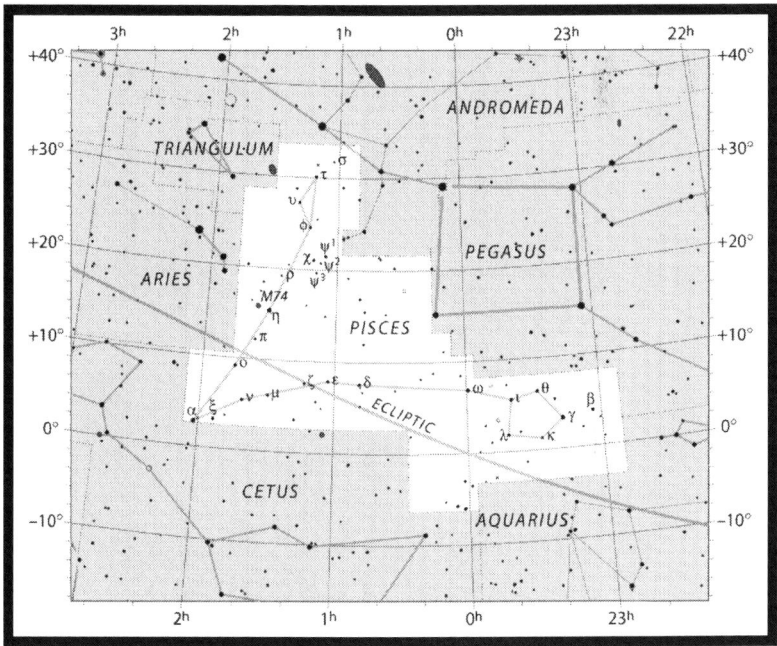

- Best observed in **Fall**
- The **Piscids** meteor shower peaks in August
- The brightest star in the constellation is **Eta Piscium**

AQUARIUS & PISCIS AUSTRINUS

THE WATER-BEARER & THE SOUTHERN FISH

Next is Aquarius, or Ganymede the "Water-bearer".

It is near Pisces and is home to many popular deep sky objects, including the **Helix nebula**, the **Saturn nebula**, and three objects from the **Messier** catalog (see table below).
Just south of Aquarius you will find Piscis Austrinus. Unlike Pisces, Piscis Austrinus represents only one fish. There are not many deep sky objects in the constellation.

NOTABLE OBJECTS	M2	M72	M73	NGC 7293	NGC 7009
TYPE	CLUSTER	CLUSTER	ASTERISM	NEBULA	NEBULA
CATEGORY	GLOBULAR	GLOBULAR	N/A	PLANETARY	PLANETARY

Aquarius pouring water into the Southern fish's mouth

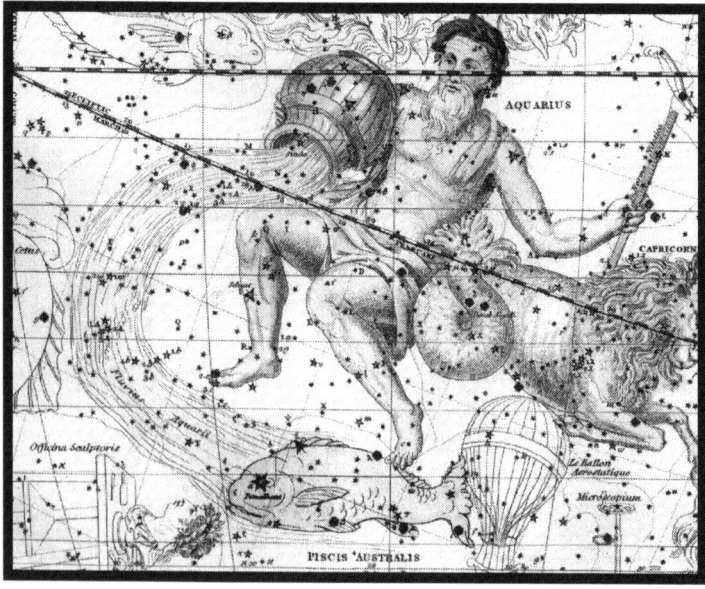

AQUILA

THE EAGLE

Ganymede's abductor, the eagle that helps carry Zeus' thunderbolts, rests in the sky as the constellation of Aquila. First described by **Eudoxus** four centuries BC, Aquila was officially published by **Ptolemy** in the 2nd century.
The constellation lies in the Milky Way band, and is visible high in the sky during the summer months.
Aquila is easy to find because its brightest star is part of the famous **Summer Triangle**. It contains mostly planetary nebulae, none of which are from the Messier catalog.

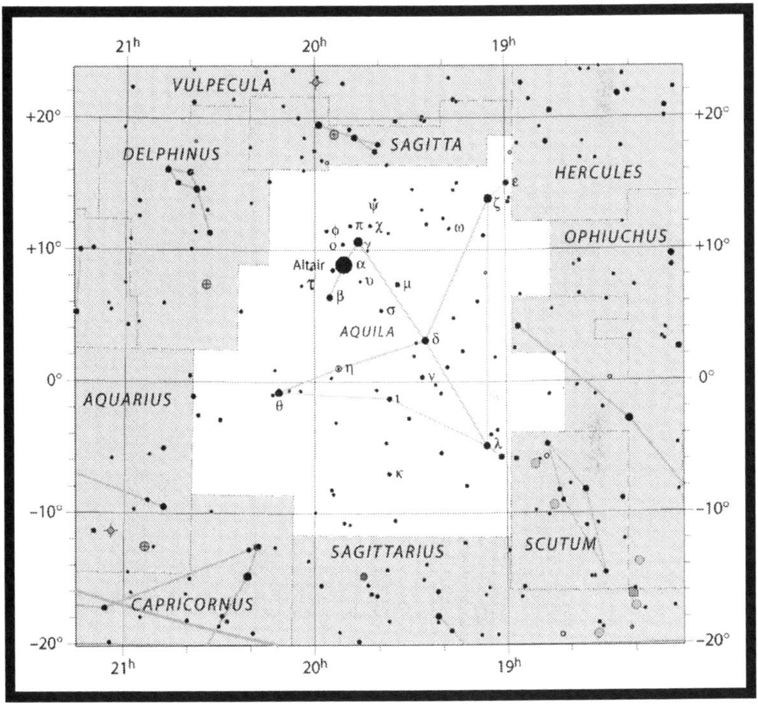

- Best observed in **Summer**
- The **June** and the **Epsilon Aquilids** occur in Aquila
- The brightest star in the constellation is **Altair**

SAGITTA

THE ARROW

Out of all 88 constellations of the night sky, Sagitta is the third smallest. Because of its size, it does not contain many deep sky objects, but it does host a Messier object, globular cluster **M71** discovered in 1746.
Sagitta is associated with the thunderbolt Zeus used to finish off Typhon, however, depending on the myth it can also represent the arrow Herakles used to kill Zeus' eagle.

Sagitta will be referenced again in the story of Herakles.

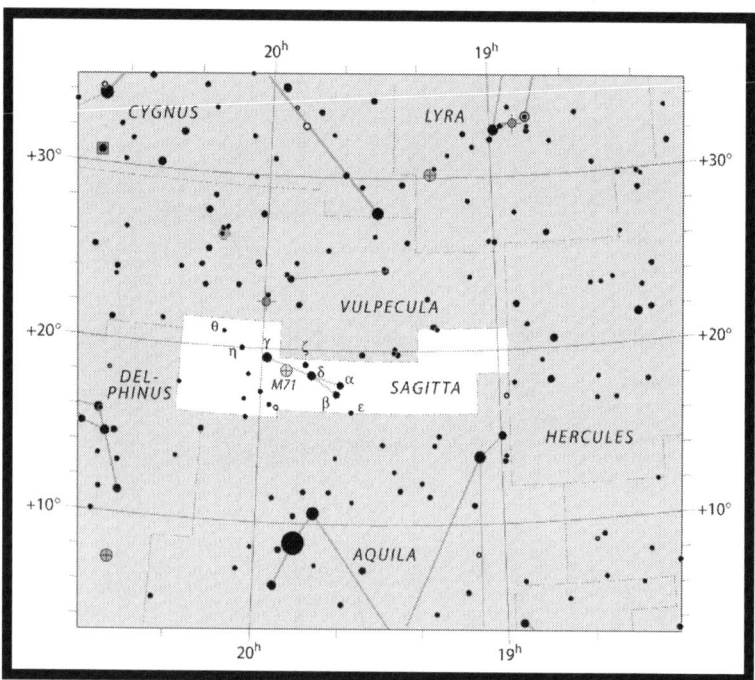

- Best observed in **Summer**
- There are **no** meteor showers associated with Sagitta
- The brightest star in the constellation is **Gamma Sagittae**

LYRA

THE LYRE

Let's get back to the water! Lyra is home to another of the three stars forming the famous **Summer Triangle**: Vega. The most famous object in Lyra is the **Ring nebula** (M57), which was the second nebula ever found.

In 7th century BC, poet and musician Arion, very skilled with the Lyre instrument became famous and wealthy while playing a tour in Italy. On the boat ride home, the sailors traveling with him became jealous and intended to kill him. Before they murdered him, Arion asked for one favor: to play a final song…

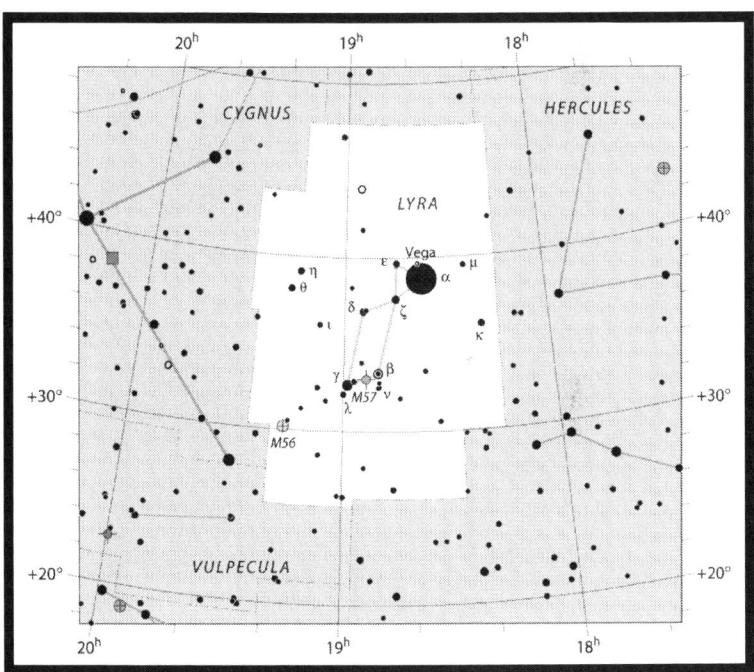

- Best observed in **Summer**
- There are **three** meteor showers associated with Lyra
- The brightest star in the constellation is **Vega**

DELPHINUS

THE DOLPHIN

…The crew agreed, and the music produced from Arion's **Lyre** attracted several **dolphins**. Instead of being killed by the sailors, Arion decided to jump into the sea. One of the dolphins saved him and helped him get back home. Zeus placed the image of a dolphin in the sky, for all the times dolphins have saved humans.

Delphinus is not far from the celestial equator. It is faint but contains several deep sky objects, although none are popular among amateur astronomers.

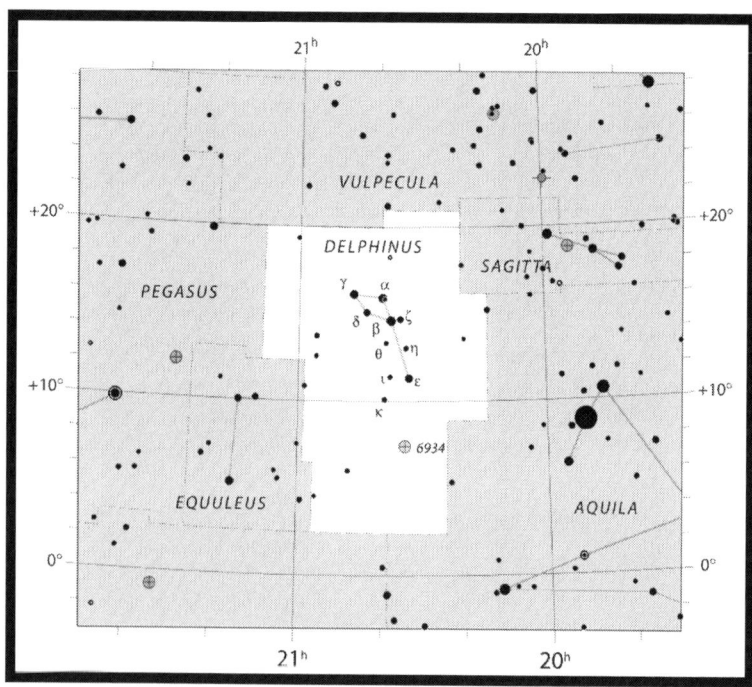

- Best observed in **Summer**
- There are **no** meteor showers associated with Delphinus
- The brightest star in the constellation is **Rotanev**

ERIDANUS

THE RIVER

Eridanus is a constellation next to the famous Orion constellation. It is mostly known for being home to the petrifying IC 2118, or the **Witch Head nebula**. The witch is illuminated by the 7th brightest star in the entire sky and the brightest in the Orion constellation: Rigel.

You can easily remember Eridanus by learning its story and how it interacts with **Cygnus** on the next page!

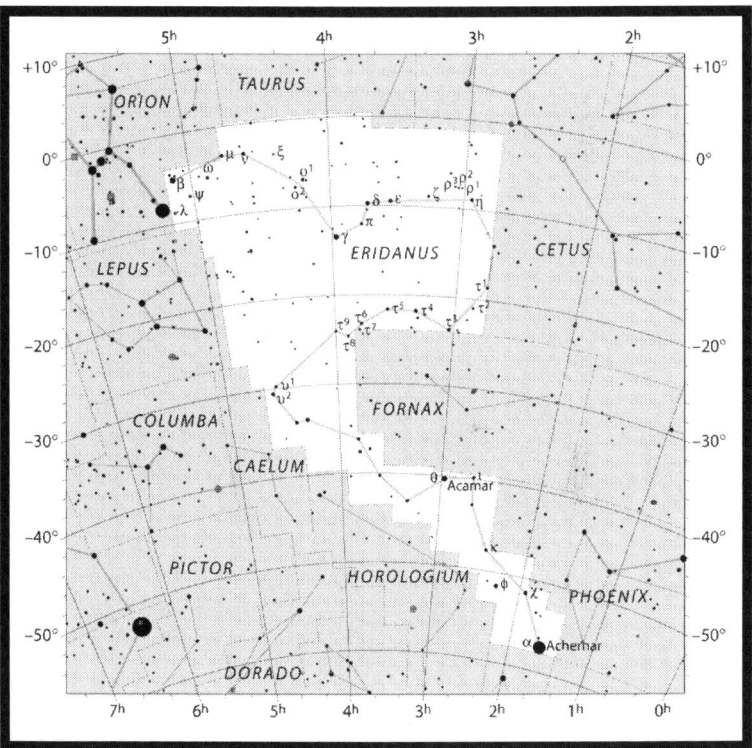

- Best observed in **Winter**
- There are **no** meteor showers associated with Eridanus
- The brightest star in the constellation is **Achernar**

CYGNUS

THE SWAN

Also called the **Northern Cross** due to its asterism, Cygnus is one of the brightest and most recognizable constellations of the night sky. Its brightest star, **Deneb**, is the last of the three stars forming the Summer Triangle along with **Altair** (from Aquila) and **Vega** (from Lyra).
Cygnus contains popular nebulae, like the **Veil**, **the North America**, and the **Pelican** nebulae. You can also spot two Messier objects, **M29** and **M39**, both being open star clusters.

THE MYTH

Phaethon, son of sun god Helios, was a stubborn teenager who wanted to ride his father's chariot so bad that he stole it. Despite his friend **Cygnus'** disapproval, Phaethon rode the chariot into the sky, but quickly realized he wasn't as strong as his father to control the two horses pulling it. The chariot flew so uncontrollably that it started to burst into flames, the trail of smoke painting the night sky with the Milky Way band. The teenager then fell to his death in the river: **Eridanus**.
Cygnus discovered his friend's body, trapped underwater, and begged zeus to transform him into a swan so that he could retrieve Phaethon and give him a proper burial. Cygnus was then placed into the night sky, in the middle of the fiery trail of smoke created by his friend's flight, the Milky Way.

The Fall of Phaethon

LOOKING UP

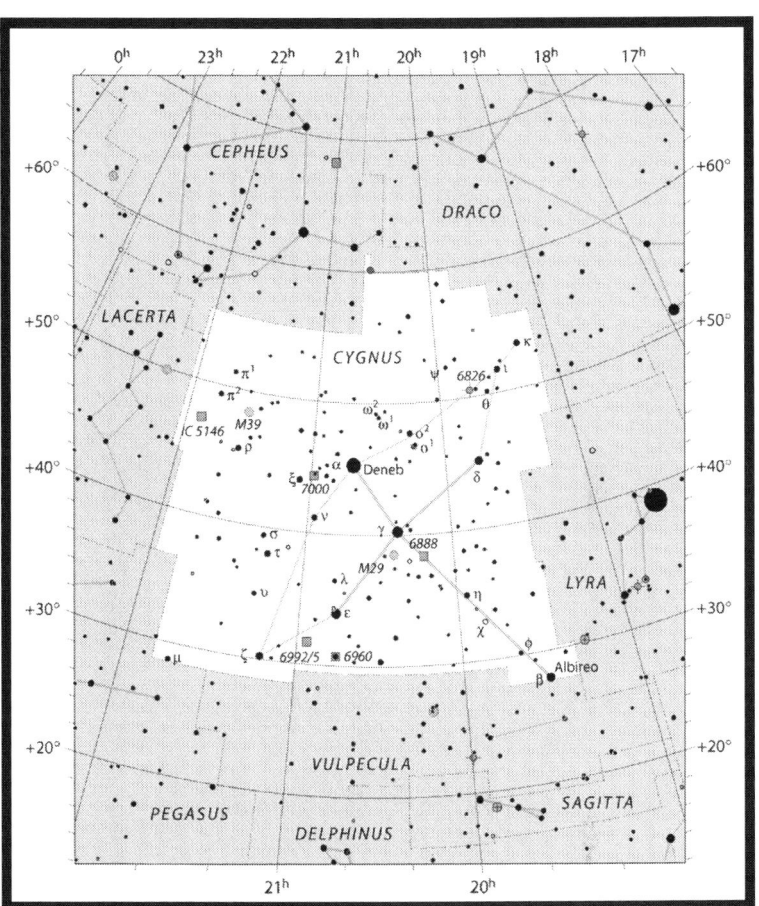

THE LOVE OF ORION

Stepping out of the daring water escapades, the next section of stories relate to the most famous hunter in Greek mythology: The Mighty Orion.

Once again, these tales will create connections to remember the stories more easily. Discover which constellation Orion was chasing until his death, and which one he was hiding from…

ORION, TOO MUCH LOVE?

Orion, giant son of Poseidon and Euryale, was the mightiest hunter of all.
Orion, thanks to his father, had the ability to walk on water and decided to visit the island of Chios. Once there, he drank too much and almost sexually assaulted Merope, one of the beautiful seven sisters. The King of the island, Oenopion, enraged at his actions, banished Orion and turned him completely blind.
The hunter, confused and unaware of his surroundings, stumbled onto another island, Lemnos. God of forging, blacksmithing and sculpture, Hephaestus, (**Auriga**) noticed the hunter, hopelessly wandering around and had pity for the man. He asked his servant, the small Cedalion, to be Orion's eyes and guide him towards the rising sun in the east. Upon reaching the sunrise, the sun god, Helios, fully restored Orion's eyesight.

The hunter is known for being in love with the **Pleiades**, the seven beautiful daughters of Atlas and Pleione. Orion would pursue them, day after day, until Zeus grabbed the sisters and placed them into the sky, so they would not have to run anymore. The Pleiades can be found in the constellation of **Taurus**, and are the most visible star cluster in the night sky.

Orion traveled to yet another island, Crete, where he met Artemis, goddess of the hunt. Orion learned a lot from Artemis and the two became hunting companions. One day, the man became so confident that he yelled he could kill any beast on the planet. Mother Nature, Gaia, heard him and was infuriated. She then sent a giant scorpion to sting and kill the hunter.
Both Orion and the animal were placed in the heavens. The constellation of the scorpion, **Scorpius**, lies at the exact opposite of Orion, and rises when the hunter sets. It is said that Orion is always fleeing from the beast responsible for his death. The constellation of the Hunter is located right behind Taurus, and can still be seen chasing the seven sisters. He will eventually manage to catch up to them as the Pleiades are slowly drifting towards the constellation of the hunter.

ORION

THE HUNTER

Located on the celestial equator, Orion is one of the most iconic constellations. It has been known for thousands of years long before Ptolemy popularized it.
Orion is a well known patch of sky amongst astronomers and astrophotographers, as it is home to several of the best nebulae in the entire night sky, like the **Orion nebula** (M42), the **Horsehead nebula** (IC434), the **Flame nebula** (NGC 2024), **Barnard's Loop**, and more!
It also contains two of the brightest, visible stars, **Rigel** and **Betelgeuse**.

The Orionids meteor shower, caused by Halley's Comet, peaks on October 21st of every year.

THE ORION NEBULA

Visible in Orion's Sword, the Orion Nebula is a very bright emission nebula that was "officially" discovered in 1611. At a distance of only 1,600 light years, it is the nearest stellar nursery to Earth.
M42 is the most popular nebula to observe through telescopes or binoculars. It is the easiest to find and is, for many people, the most impressive object to view. It is visible with the naked eye as well, although it looks just like a star.

Separated only by a dust lane is **M43**, or De Mairan's nebula, which looks like it is part of M42. With a telescope, you can also spot the **"Trapezium"**, an open star cluster that illuminated all the gases around it.

LOOKING UP

- Best observed in **Winter**
- There are **two** meteor showers associated with Orion
- The brightest star in the constellation is **Rigel**

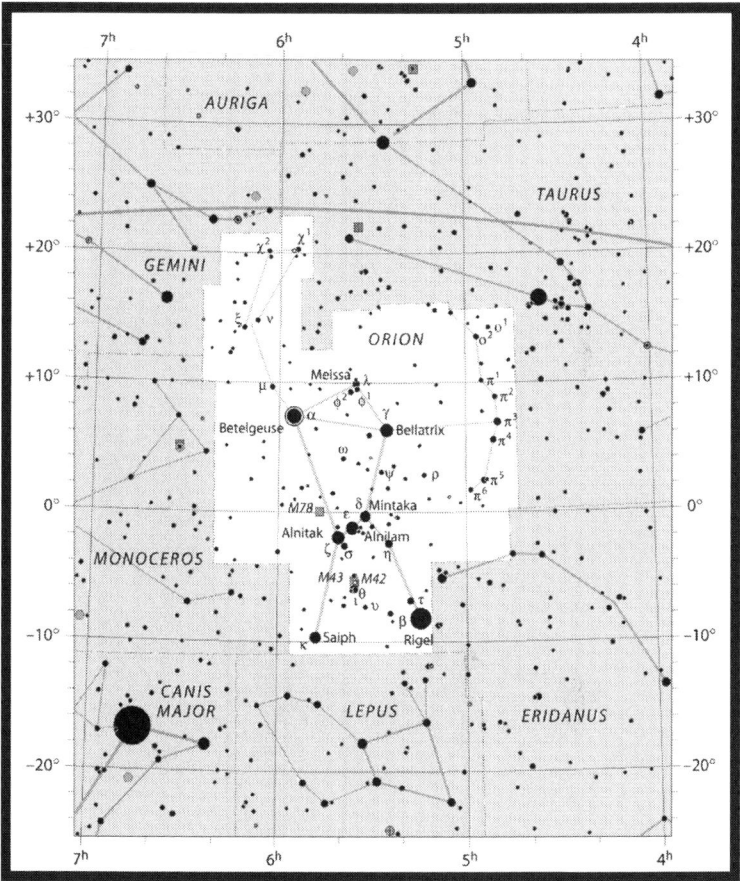

NOTABLE OBJECTS	M42	M43	M78	IC 434	NGC 2024
TYPE	NEBULA	NEBULA	NEBULA	NEBULA	NEBULA
CATEGORY	EMISSION	EMISSION	REFLECTION	DARK	EMISSION

CANIS MAJOR & LEPUS

THE GREATER DOG AND THE HARE

Canis Major and Lepus are visible constellations following the Hunter. Canis Major is easy to find, as it hosts the brightest star in the entire night sky: **Sirius** ("The Dog Star").

Several deep sky objects can be found in Canis Major, such as the open cluster **M41**, and the beautiful emission nebula NGC 2359, also known as **Thor's Helmet**. In 2003, astronomers discovered the closest galaxy to the Milky Way in the constellation of the greater dog, and named it the **Canis Major Dwarf Galaxy**.

The only deep sky object in Lepus is the globular cluster **Messier 79.**
In Greek mythology, Canis Major, represents Orion's big dog and can be seen hunting the hare (Lepus).

Canis Major hunting Lepus

CANIS MINOR

THE LESSER DOG

Canis Minor, or "The Lesser Dog" represents Orion's smaller dog, also hunting the hare with **Canis Major**.

Canis Minor only has two stars that are bright enough to be visible easily with the naked eye.
Although the Milky Way band passes through Canis Minor, the constellation does not contain any interesting objects for amateur astronomers. Its brightest star is **NGC 2485**, a spiral galaxy with a magnitude of 12.4.

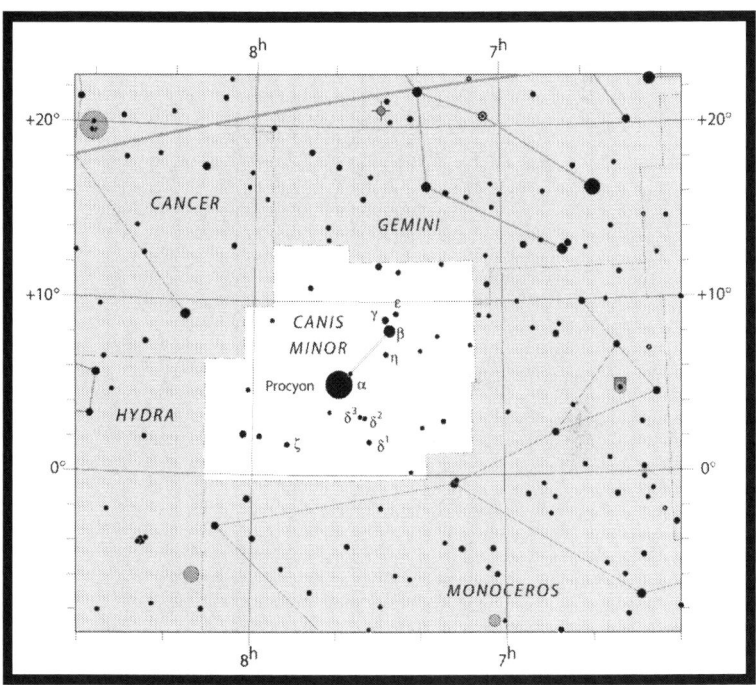

- Best observed in **Winter**
- The **Canis-Minorids** meteor shower occurs in Canis Minor
- The brightest star in the constellation is **Procyon**

TAURUS

THE BULL

Taurus is a constellation that represents a bull, often said to be fighting Orion.

It is home to the two nearest open star clusters to Earth: the famous **Pleiades** (M45), and the **Hyades.** It also contains a popular target for amateur astronomers and astrophotographers, the **Crab nebula**, which was the very first object entered into Charles Messier's catalog.

Taurus is the 17th largest constellation in the night sky, and is one of the brightest overall. Its brightest star, **Aldebaran**, is the 13th brightest in the sky and is a beautiful, orange giant with a magnitude of -2.1.

There are two meteor showers associated with Taurus, the **Taurids,** which peak in November, and the **Beta Taurids,** which occur in June and July.

THE SEVEN SISTERS

The constellation of Taurus and the Messier 45 star cluster, have been known since before 15,000 BC. Paintings of the bull and the Pleiades have been found in the **Lascaux caves** near the south of France.

Can you spot the seven sisters? The dots above the bull represent the **Pleiades**, while the dots on the bull's face depict the other star cluster in Taurus: The **Hyades**.

LOOKING UP

Taurus is best observed in **Winter**, and can be found in between Orion, Auriga, Perseus, Aries and Eridanus.

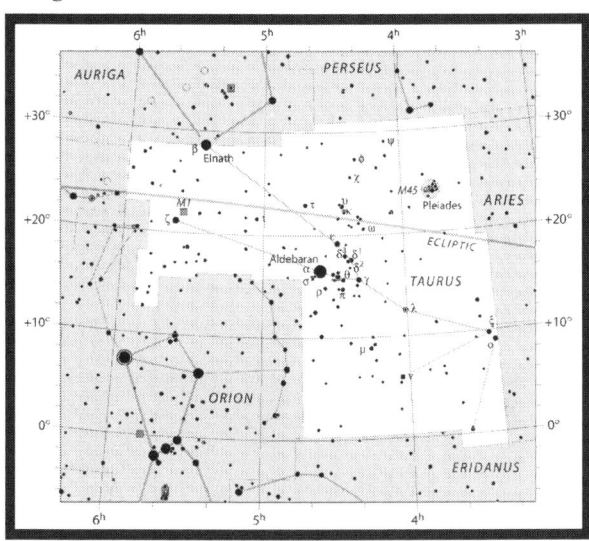

The Seven Sisters, painting from 1885

AURIGA

THE CHARIOTEER

Capella, the 6th brightest star in the sky, is one of the stars forming the constellation of Auriga. In Greek mythology, the name is Latin for "the Charioteer" and is thought to represent **Hephaestus** himself. Hephaestus was the god of forging, blacksmithing and sculpture, and, inspired by **Helios'** chariot, built his own so he could go anywhere. Hephaestus also helped Orion regain his sight.

There are three open clusters in Auriga, **M36**, **M37**, and **M38**, as well as a nebula, IC 405, also known as the **Flaming Star nebula**.

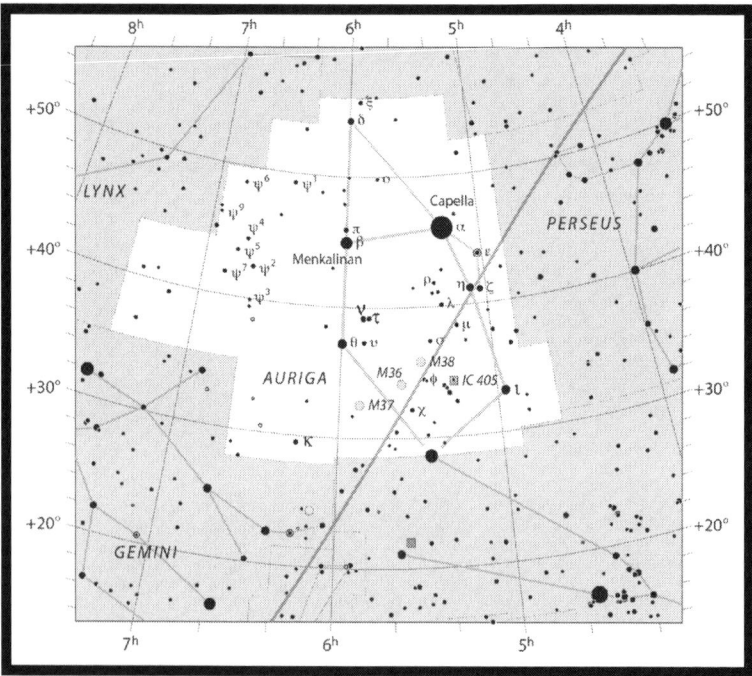

- Best observed in **Winter**
- There are **two** meteor showers associated with Auriga
- The brightest star in the constellation is **Capella**

GEMINI

THE TWINS

Next to Auriga is another great constellation. Gemini is part of several myths, but none of them are linked to Orion's story. Ptolemy associated **Pollux** and **Castor**; the two brightest stars that form the head of each twin, with Apollo and Herakles, half-brothers and both sons of Zeus.

Gemini contains several interesting targets, such as the open cluster **Messier 35**, the **Eskimo nebula**, the **Jellyfish nebula**, and more! It also has a neutron star, by the name of Gaminga, meaning "It's not there".

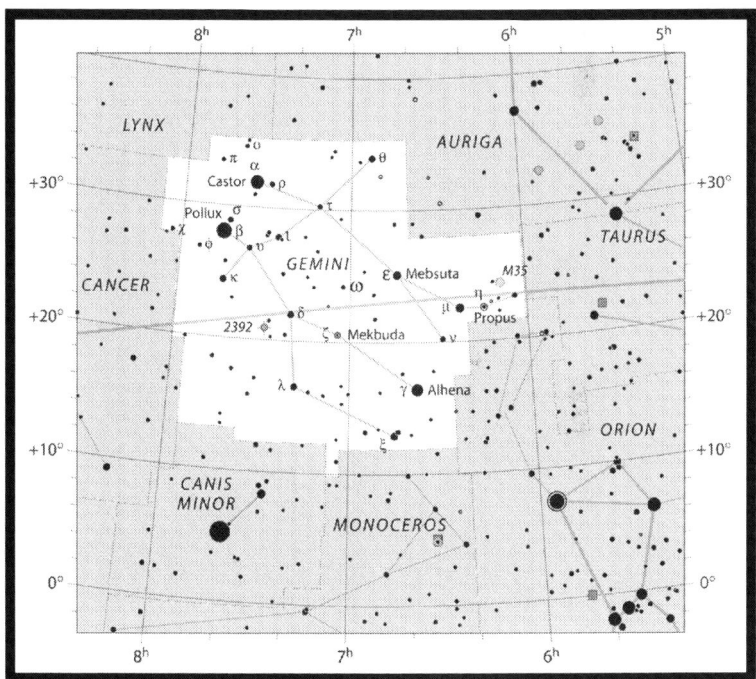

- Best observed in **Winter**
- There are **two** meteor showers associated with Gemini
- The brightest star in the constellation is **Pollux**

SCORPIUS

THE SCORPION

Scorpius... The beast that killed Orion. The two of them can never be seen at the same time, because as one sets in the West, the other rises in the East.

The main feature of the constellation is its brightest star, the red supergiant **Antares**. In Greek mythology, Antares represents the heart of the scorpion and is so red that it looks just like Mars! It is also the 16th brightest star in the sky.

DEEP SKY OBJECTS

Many deep sky objects can be spotted in Scorpius, including the **Cat's Paw** nebula (NGC 6334), the **Butterfly** nebula (NGC 6302) and several Messier objects such as **M4**, **M6** (Butterfly cluster), **M7** (Ptolemy cluster), and **M80**.

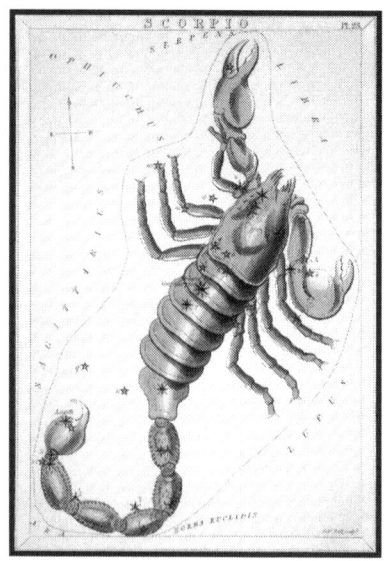

Messier 7 is nicknamed the Ptolemy cluster because the Greek astronomer took note of it in the year 130, thinking it was a nebula.

Rho Ophiuchi can be found in the constellation of Ophiuchus. The multiple star system is easy to find using Scorpius. Antares is part of the system, thus aiming a telescope or camera to the red supergiant assures you will be right on target.

LOOKING UP

The constellation of Scorpius is the 33rd largest. The scorpion is located near the center of the Milky Way, between Sagittarius and Libra.

- Best observed in **Summer**
- There are **two** meteor showers associated with Scorpius
- The brightest star in the constellation is **Antares**

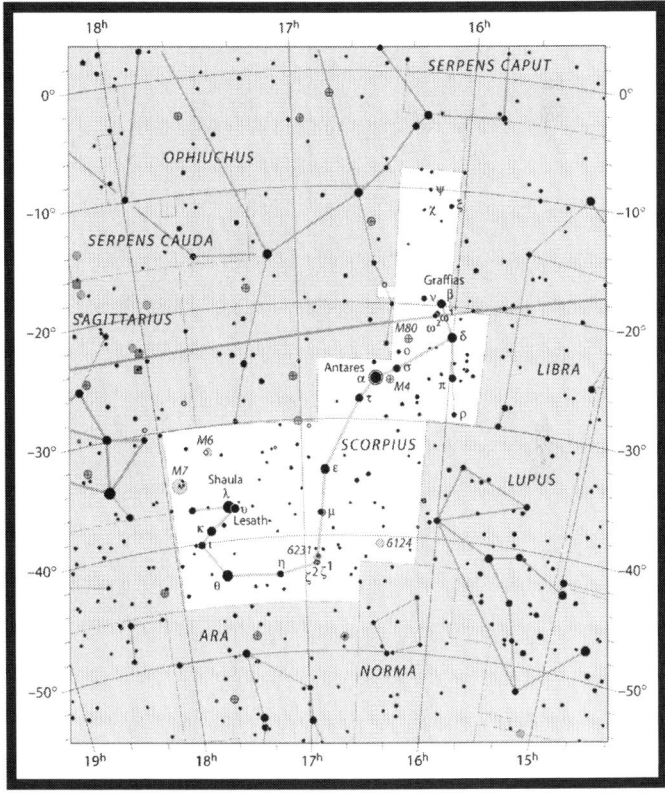

SAGITTARIUS

THE ARCHER

Sagittarius is an interesting constellation located near the galactic core and it is packed with deep sky objects! The three most popular ones observed and photographed in the constellation are M8 (**The Lagoon nebula**), Messier 17 (**The Omega nebula**), and Messier 20 (**The Trifid nebula**), but there are a total of 15 Messier objects in Sagittarius alone!

It is also home to other interesting members, such as the radio source **Sagittarius A** and the **Pistol** Star, which is about four million times brighter than our sun and almost 200 times more massive. It is one of the most luminous stars ever discovered.

The bright stars in Sagittarius also form a famous asterism: the **Teapot**. The 8 stars form the bow and arrow of the archer.

- Best observed in **Summer**
- There are **no** meteor showers associated with Sagittarius
- The brightest star in the constellation is **Kaus Australis**

ORION AVENGED?

Sagittarius represents the centaur (half human, half horse) Crotus, son of Pan (**Capricornus**).

Crotus is known for having invented archery, and can be seen in the night sky holding his bow and arrow directly towards the red supergiant Antares, representing **Scorpius**' heart.

Will Crotus ever avenge Orion by killing the scorpion? We will never know. One thing is sure, he only needs to fire that arrow to impale the beast right through its heart.

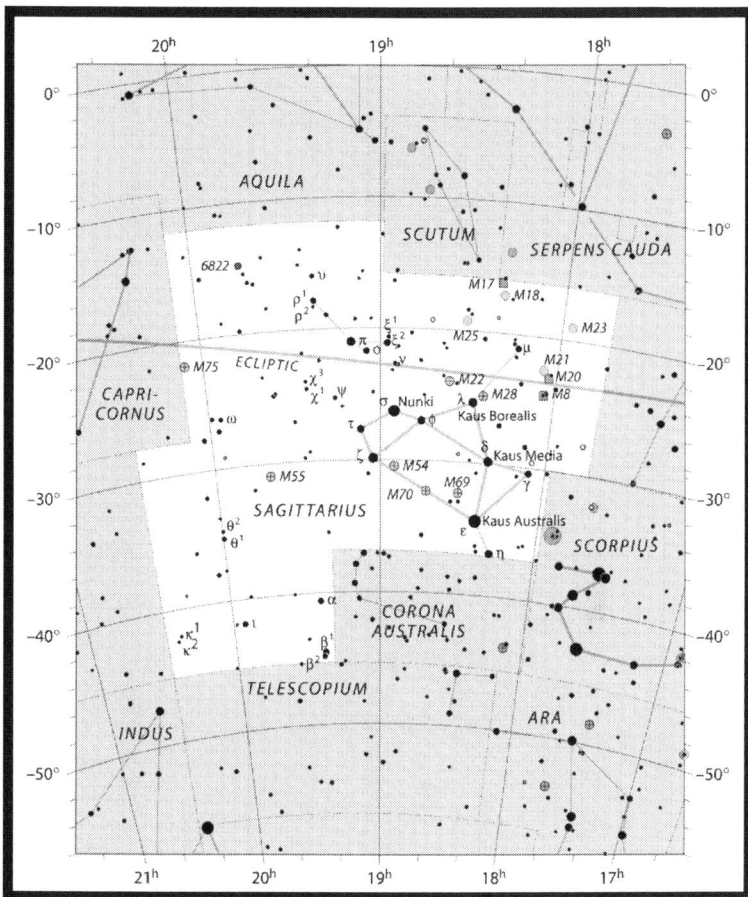

PAIRING UP

It is with this cliffhanger that the tales related to the mighty hunter, Orion, end.

The few constellations that follow do not have memorable stories attached to them, but the best way to remember them are by pairs. The next 6 constellations will go two by two.

ARIES

THE RAM

Aries represents a golden ram, and is a medium-size constellation located between **Pisces** and **Taurus**.

Three bright stars, Alpha, Beta, and Gamma Arietis, form an asterism designated by **Johann Bayer** that is often used by navigators. The constellation contains several double stars, as well as deep sky objects, mainly NGC galaxies.

Aries is home to the Daytime Arietids, one of the best meteor showers that occur during the day, peaking in the month of June.

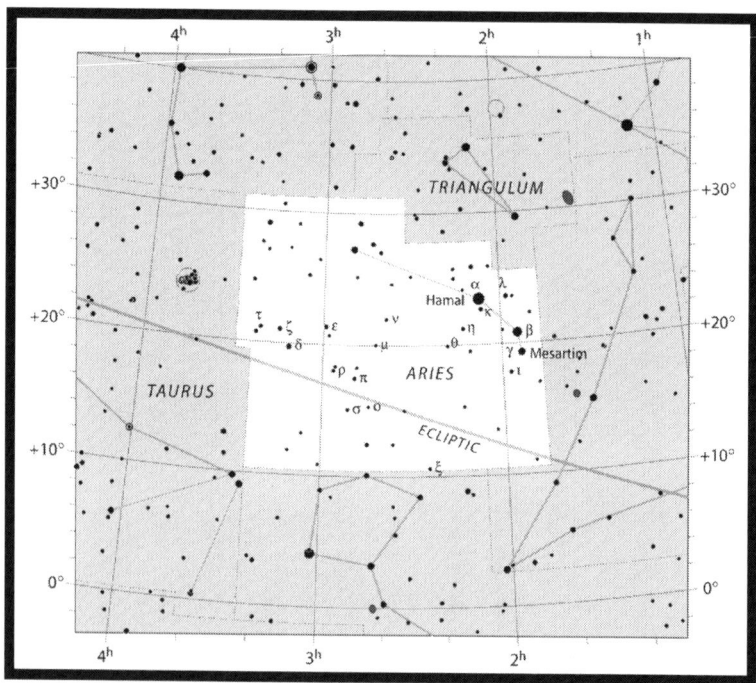

- Best observed in **Fall**
- There are **six** meteor showers associated with Aries
- The brightest star in the constellation is **Hamal**

TRIANGULUM

THE TRIANGLE

Triangulum is a very small constellation next to **Andromeda**. The Greeks called the pattern of stars "Deltoton" because it looked like the Greek letter Delta.
Although Triangulum is a tiny constellation, it hosts one of the brightest and largest visible galaxies in the northern hemisphere, the **Triangulum galaxy**, or Messier 33.
One of the most impressive features of M33 is that an amateur astrophotographer may see the emission nebula **NGC 604** in it, which is one of the most massive and brightest nebulae in the entire local group of galaxies!

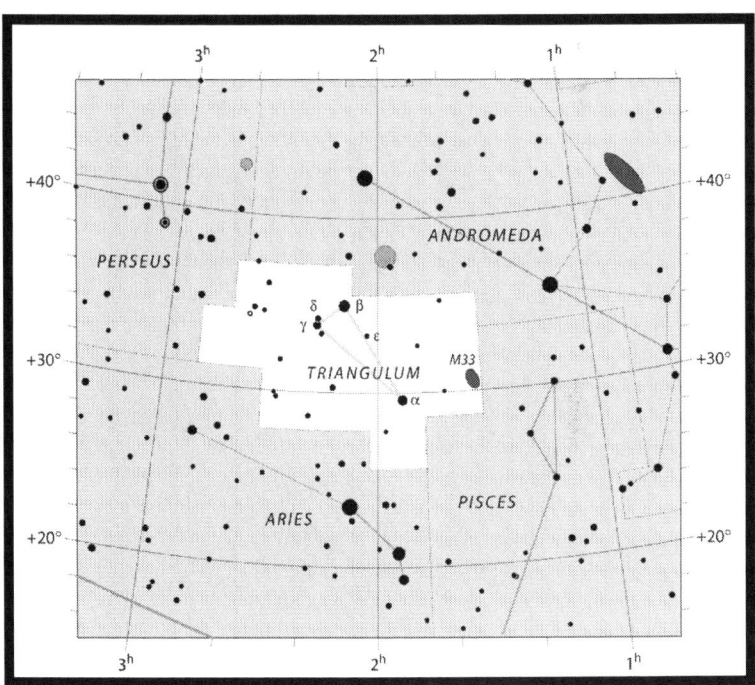

- Best observed in **Fall**
- There are **no** meteor showers associated with Triangulum
- The brightest star in the constellation is **Beta Trianguli**

OPHIUCHUS & SERPENS

THE SERPENT-BEARER AND THE SNAKE

Ophiuchus is Latin for "Serpent-bearer" and represents Asclepius, the healer in the Greek mythology. Asclepius can be seen in the sky holding a snake in his hands, represented by Serpens.

Serpens is the only constellation in the entire night sky to be located in two completely different patches of sky, with the head of the animal on one side and its tail on the other.

There are many deep sky objects in Ophiuchus, including seven from the Messier catalog! It contains the famous **Rho Ophiuchi** molecular cloud complex, seen surrounding the bright, orange star, Antares. In the tail of Serpens lies the famous **Eagle nebula** that contains the incredible Pillars of Creation.

Ophiuchus holding Serpens

CORONA BOREALIS

THE NORTHERN CROWN

One of the easiest ways to find Serpens' head is by spotting the small constellation of the Northern Crown: Corona Borealis.
The main stars of Corona Borealis form the shape of a crown, representing the one thrown into the sky by the Greek goddess Ariadne, during her wedding. It is said that the jewels from the crown, made by the god Hephaestus (**Auriga**), turned into stars to make the constellation of Corona Borealis.
The only interesting deep sky object within the Northern Crown is **Abel 2065**, a galaxy cluster containing more than 400 objects.

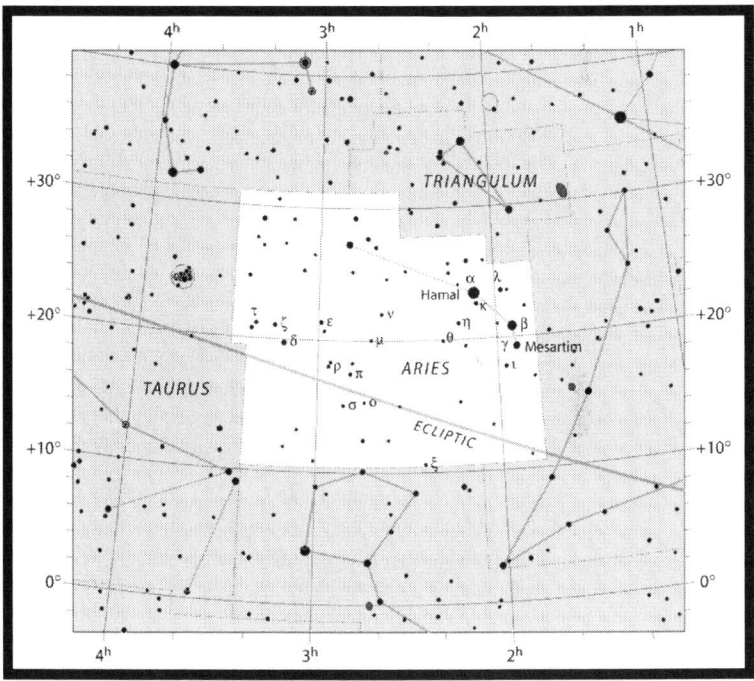

- Best observed in **Summer**
- There are **no** meteor showers in Corona Borealis
- The brightest star in the constellation is **Alphecca**

CORONA AUSTRALIS

THE SOUTHERN CROWN

Corona Autralis is Latin for "Southern Crown" and is located between the constellations of **Sagittarius** and **Scorpius**.

Corona Australis is similar in shape and in size to the Northern Crown (**Corona Borealis**) but is not associated with any known Greek myth.

There are several deep sky objects within the constellations, such as the **Corona Australis nebula**, but the rest are mostly star clusters.

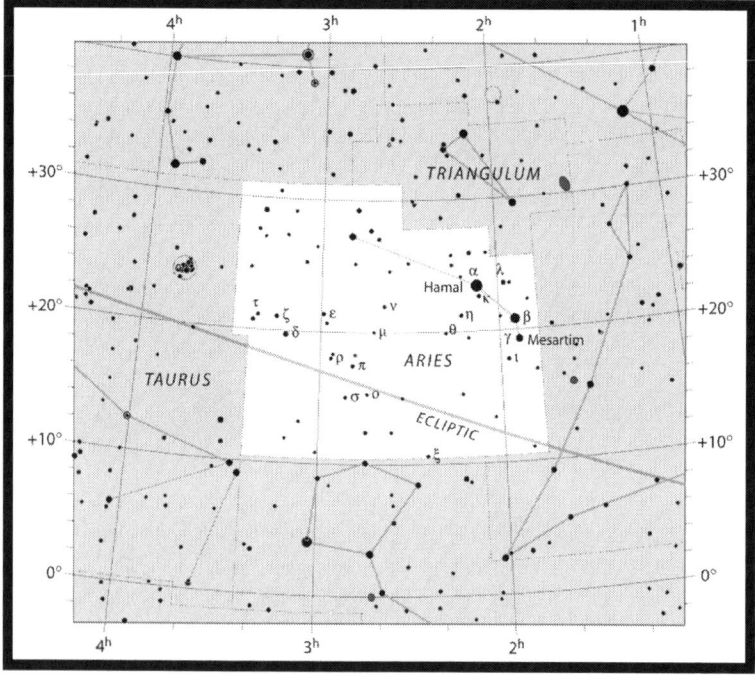

- Best observed in **Summer**
- The **Corona Austrinids** happen in Corona Australis
- The brightest star in the constellation is **Alphekka Meridiana**

VIRGO & LIBRA

THE GODDESS OF JUSTICE AND HER SCALES

Virgo and Libra are two constellations that lie next to one another. Virgo represents the Greek goddess of justice Dike and is the second largest constellation in the sky. Dike can be seen holding the scales of justice, represented by the constellation of Libra. Libra used to be part of Scorpius before it became associated with balance, and its brightest star, Zubeneschamali, means "The Northern Claw" in Arabic.

Virgo is home to eleven Messier objects, and many NGC galaxies! It is also home to the famous Virgo Cluster. Libra, on the other hand, does not contain any Messier objects, but does have a few NGC galaxies.

Virgo, with Libra to the bottom left

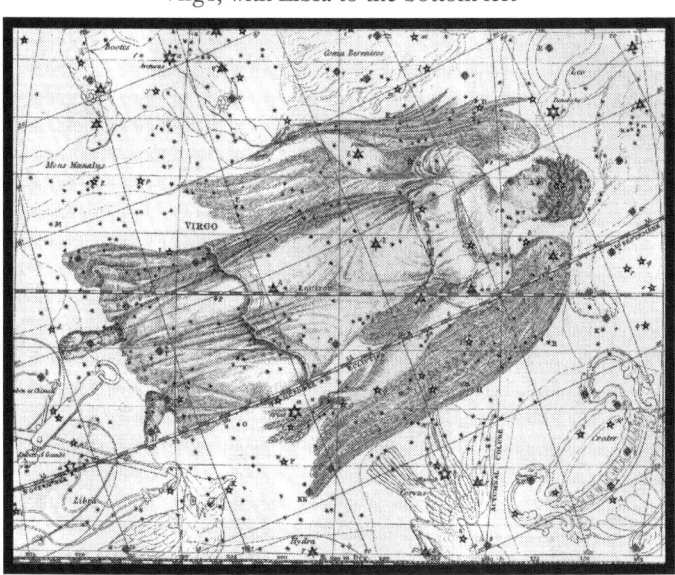

HERCULES AND THE TWELVE LABORS

Just like **Andromeda** and **Orion**, there are several constellations related to Hercules. Before discussing Greek mythology's most popular character, we begin with the story of the crow as it will link to Hercules.

One day, Apollo needed water to complete a sacrifice at the altar, **Ara**. He sent **Corvus**, his sacred crow (or raven), to fetch water in a cup, **Crater**. Obviously, there is no photo evidence, but long ago, crows were believed to have white feathers. Upon finding a water source, the bird became distracted by a fig tree and spent days eating and resting instead of fulfilling his task. Afraid of being punished for his incompetence, Corvus went back to Apollo with a fictitious story about being attacked by **Hydra**, a water snake.

The god, furious, put a spell on the bird that burned his feathers and made him perpetually thirsty. He then sent both the animal and the cup into the sky. Corvus' actions are the reason why crows and ravens have black feathers and a raspy voice.

Herakles, son of Zeus, who we refer to as "**Hercules**" due to the constellation's given name, had a rough life from birth to death but inspired thousands. Three constellations are directly linked to the famous Hercules and his 12 labors, the second one involving Hydra, the water snake.

Hydra was the spawn of a monster mentioned earlier in this book, **Typhon** (see page 25), and took the form of a giant creature with nine heads. Hercules could not kill the beast alone, as cutting off one of Hydra's heads would grow two new ones. To make matters worse, **Hera**, his worst enemy, sent a large crab called Karkinos to distract Hercules so he would fail in battle. The hero easily defeated the crab, and managed to kill Hydra with the help of his nephew, by cauterizing the headless necks before two new heads grew back.

Hydra can be seen in the sky next to Corvus and Crater, making sure that the bird does not drink from the Cup.

Hera placed the crab in the heavens for his efforts, and is represented as the constellation of **Cancer**.

CORVUS

THE CROW

Corvus is a small constellation that can be seen by observers in both hemispheres.
There are no notable objects in Corvus besides the **Antennae Galaxies**. NGC 4038 and NGC 4039 are two colliding galaxies that have been merging for the past 600 million years. The two cores are expected to become one in about 400 million years. Their collision was so intense that the gas expelled by each galaxy formed the shape of antennae, hence the name. This is what we may expect to "see" when our Milky Way galaxy will collide with the Andromeda galaxy.

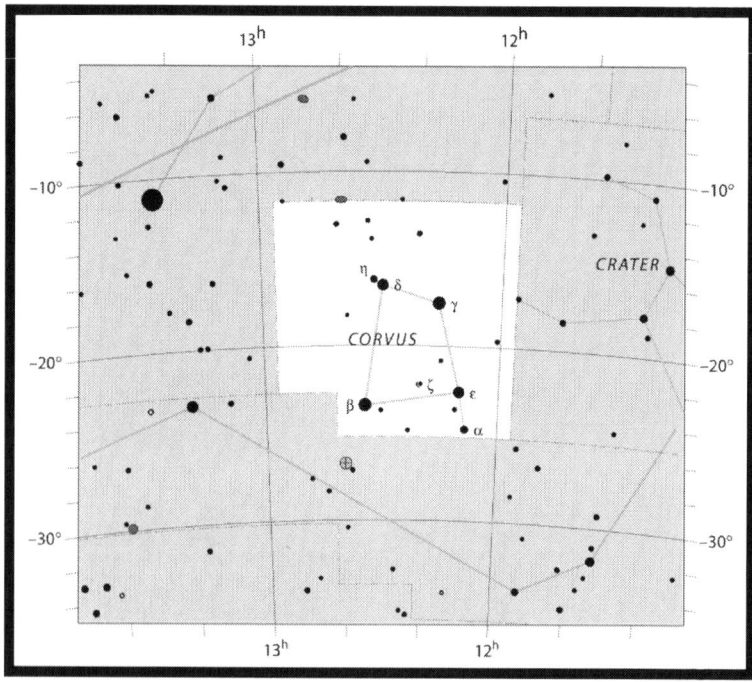

- Best observed in **Spring**
- There are **two** meteor showers associated with Corvus
- The brightest star in the constellation is **Gienah**

CRATER

THE CUP

Another southern constellation next to **Corvus** and **Hydra** is Crater, the cup.

Crater is also small and contains stars that are no brighter than 3.56 in magnitude. There are no Messier objects in the constellation nor any interesting galaxies, nebulae, or star clusters for amateur astronomers and astrophotographers to view.

Crater is, without a doubt, one of the most uninteresting constellations in the entire night sky.

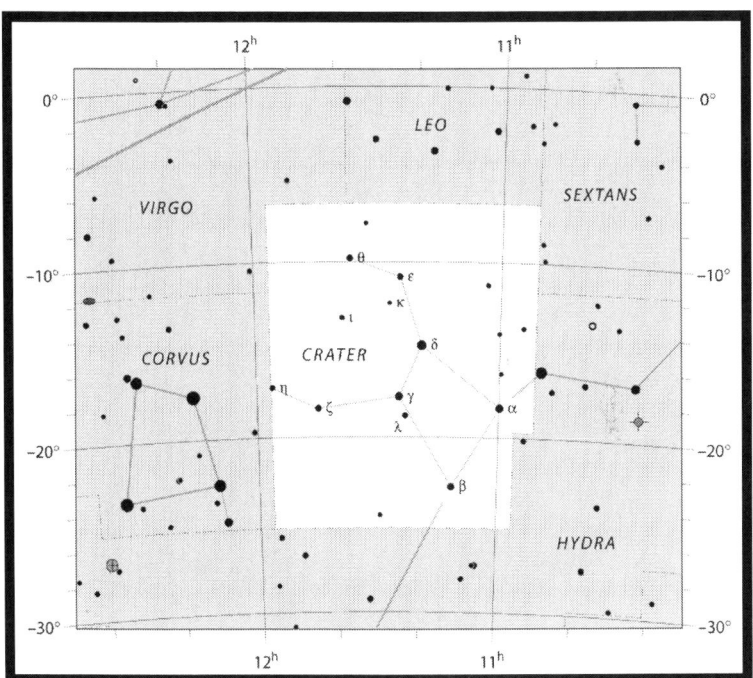

- Best observed in **Spring**
- The **Eta Craterids** happen in Crater
- The brightest star in the constellation is **Delta Crateris**

HERCULES

SON OF ZEUS

Hercules is located between Lyra and Corona Borealis. It is the fifth largest constellation in the night sky and is one of the oldest named pattern of stars known.

Hercules contains two globular clusters from the Messier catalog: **M13** and **M92**. Several other objects, mostly galaxies, are visible in the constellation, but are difficult to admire with amateur equipment.

The **Hercules-Corona Borealis Great Wall** is also partly located in Hercules. It is the largest galactic superstructure in the entire universe.

THE GREAT GLOBULAR CLUSTER

Messier 13 is the brightest and most visually impressive globular cluster in the northern hemisphere.

At half a million solar masses, the cluster is made up of more than **500,000** stars, and is one of the **oldest** observable objects in the night sky.

In 1974, M13 was the target used for sending a message to potential extraterrestrial civilizations. Sadly, we will not be alive to see the fruit of our efforts. It will take about **25,000** years before the message comes close to its vicinity, plus the same amount of time for a potential answer to return!

LOOKING UP

- Best observed in **Summer**
- The **Tau Herculids** happen in Hercules
- The brightest star in the constellation is **Kornephoros**

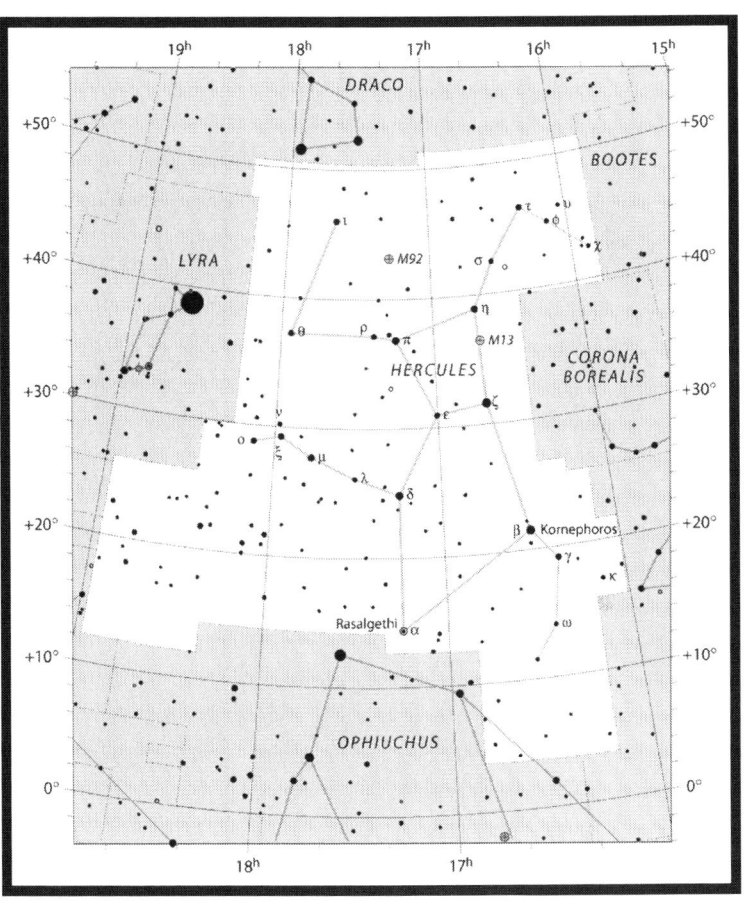

CANCER

THE CRAB

Not far from Hydra and Leo is Cancer, a medium-size constellation.

Its name is Latin for "The Crab" and is a representation of Karkinos, the crab sent by Hera to distract Hercules during his fight against **Hydra**.
Cancer contains one of the most famous star clusters in the night sky, **Messier 44**, also called the **Beehive cluster**. The only other interesting object for amateur astronomers in Cancer is another from the Messier catalog: the open cluster **M67**.

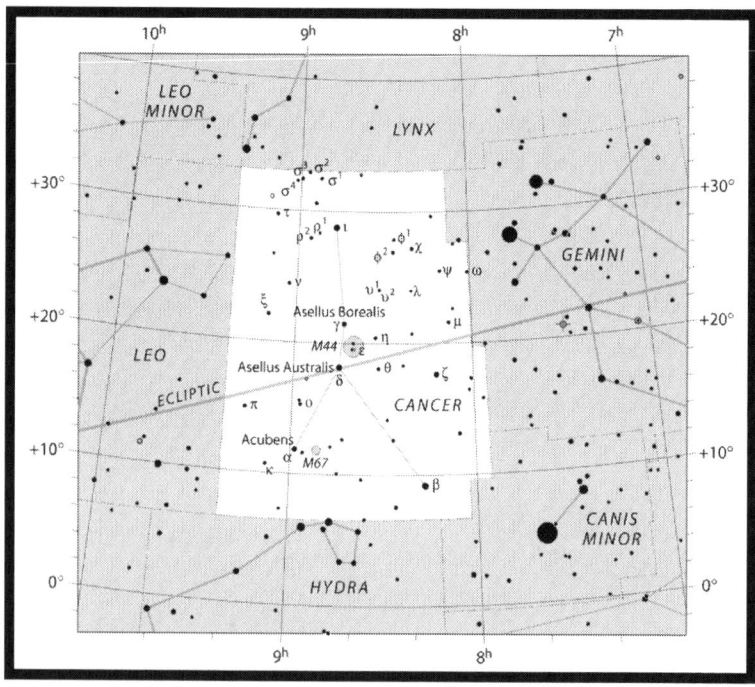

- Best observed in **Winter**
- The **Delta Cancrids** happen in Cancer
- The brightest star in the constellation is **Beta Cancri**

HYDRA

THE WATER SNAKE

Hydra's head is near **Cancer** and stretches far as its tail goes all the way to **Centaurus**. Covering more than 1300 square degrees, Hydra is the largest constellation in the entire night sky.

Due to its size, it is home to several interesting deep sky objects, including three from the Messier catalog: the open star cluster **M48**, the globular cluster **M68**, and **M83**, also called the **Southern Pinwheel galaxy** due to its ressemblance to M101.

Other notable objects within the constellation would be **NGC 3242** (The Ghost of Jupiter) and the **Hydra cluster** of galaxies. Other deep sky objects visible in the water snake are mostly NGC galaxies.

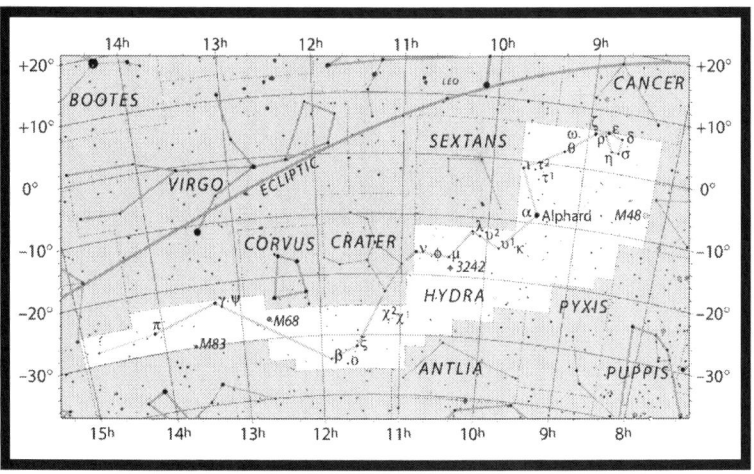

- Best observed in **Spring**
- There are **two** meteor showers associated with Hydra
- The brightest star in the constellation is **Alphard**

THE LIFE AND DEATH OF CHIRON

After killing **Hydra**, Hercules dipped his arrow into the poisonous blood of the monster, so that he could kill his future enemies with ease. However, this action will eventually lead to the hero's accidental death.

Chiron was Hercules' hunting teacher and friend. He was also a devoted medicine researcher. Half-man, half-horse, Chiron was the only immortal centaur there was, and is represented by the constellation of **Centaurus**.
Next to Chiron is **Lupus**, the wolf, who is seen being impaled by the avid hunter, as a sacrifice to the altar, **Ara**.
Chiron had a daughter, Hippe. One day, she became pregnant, but was too afraid to tell her father. She hid for months, deep in the mountains, until her child was born. Chiron, worried and furious at her actions kept looking for her, but was unable to find her as the gods turned her into a mare. She can be seen in the sky as **Equuleus**, still hiding from her father, with only her head visible behind the flying horse **Pegasus**.

As for Chiron, he suffered an accidental death by Hercules' poisoned arrows. Unaware of a battle involving Hercules in the Centaur's hometown, a stray arrow struck his knee mid-fight. Hercules, realizing his mistake, tried everything he could to heal his friend and mentor, but Hydra's poison had no cure. Sadly, the centaur was immortal and could not die, thus was doomed to live an eternal life of terrible suffering and agony. Zeus eventually freed Chiron from this fate by allowing the kind centaur to die. He then placed him in the heavens.

As you read earlier, there are three constellations associated with Hercules' twelve labors, one being Hydra.
Another is **Leo**, who represents the Lion with the impenetrable skin that Hercules was forced to strangle in order to kill. Successful in his feat he wore its fur as a cloak.
The last is **Draco**, the dragon guarding the golden apples on Mount Atlas that Hercules had to steal for his 11th task.

CENTAURUS

THE CENTAUR

Centaurus is one of the most interesting constellations in the night sky. It is very large and easily recognizable as it contains two of the brightest stars, **Alpha Centauri**, and **Beta Centauri**.

Not only is it home to the famous **Omega Centauri** globular cluster, it also contains the very bright galaxy **Centaurus A**, the **Blue Planetary** NGC 3918, and more!

You can also spot one of the largest stars known: **V766 Centauri**, and the closest star to ours: **Proxima Centauri**.

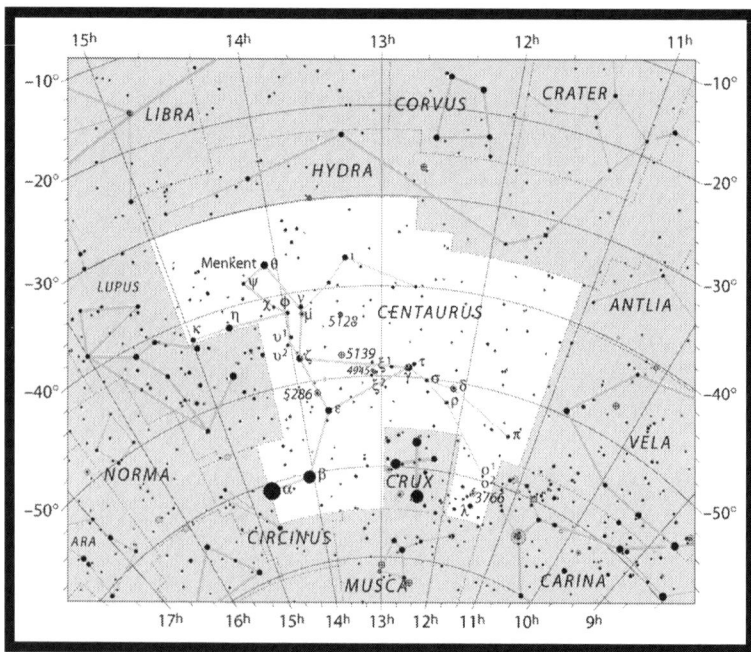

- Best observed in **Spring**
- There are **three** meteor showers associated with Centaurus
- The brightest star in the constellation is **Alpha Centauri**

LUPUS & ARA

THE WOLF AND THE ALTAR

Lupus and Ara are two small constellations next to **Centaurus**.
Lupus, the wolf, can be seen being impaled by Chiron, and given to **Ara**, the Altar, as a sacrifice.

There are several deep sky objects in Lupus although none of them are easily seen using amateur equipment. **Alpha Lupi** is the brightest star in the constellation, shining at a magnitude of 2.3. For comparison, the brightest one in Ara is **Beta Arae**, with a magnitude of 2.8.
Ara occupies an area of 0.575% of the night sky, making it the 63rd largest constellation by size.

Centaurus (left), Lupus (center) and Ara (right)

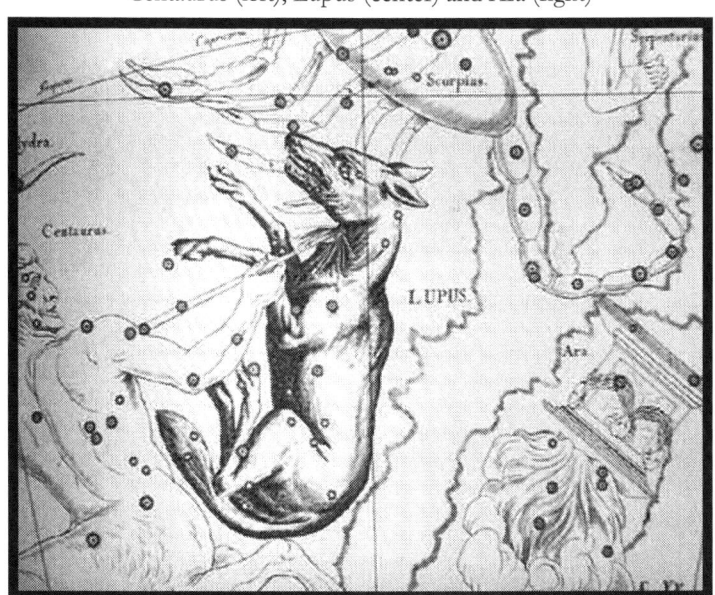

EQUULEUS

THE PONY

Equuleus is Latin for "Pony", and it is one of the least interesting constellations in the entire night sky. It is very small (the second smallest after **Crux**), faint, and does not contain any impressive deep sky objects.

Equuleus can be found next to **Pegasus**. It is sometimes called "The First Horse" as it rises just before the other.

Some objects found in the constellation include **NGC 7015**, **NGC 7045** and **NGC 7046**, however, they are all but impressive.

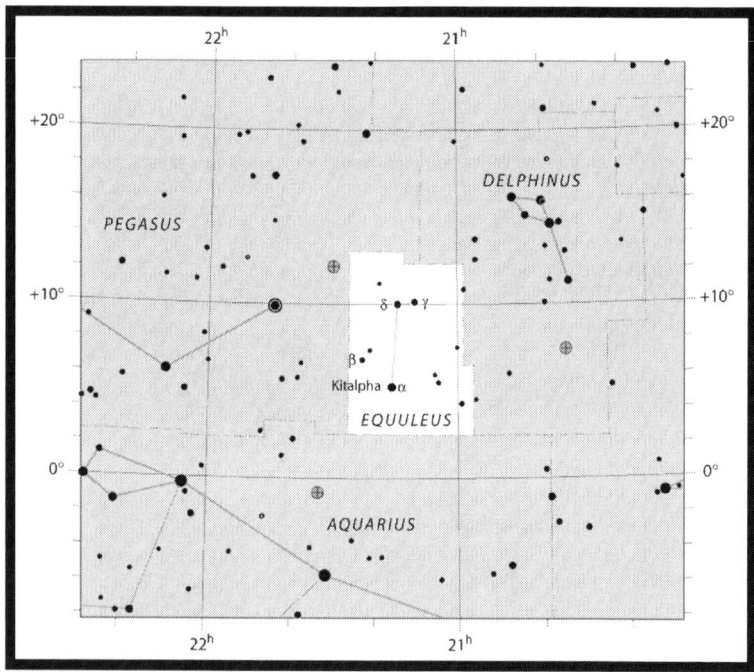

- Best observed in **Fall**
- There are **no** meteor showers associated with Equuleus
- The brightest star in the constellation is **Kitalpha**

LEO

THE LION

In between **Cancer** and **Virgo** is the constellation of the lion: Leo. Its stars are bright, making it recognizable and lies just above the ecliptic line.

There are 5 Messier galaxies in Leo, two of them (**M65** and **M66**) being part of the famous **Leo Triplet**. The three others, **M95**, **M96** and **M105** are part of the Leo I group of galaxies.

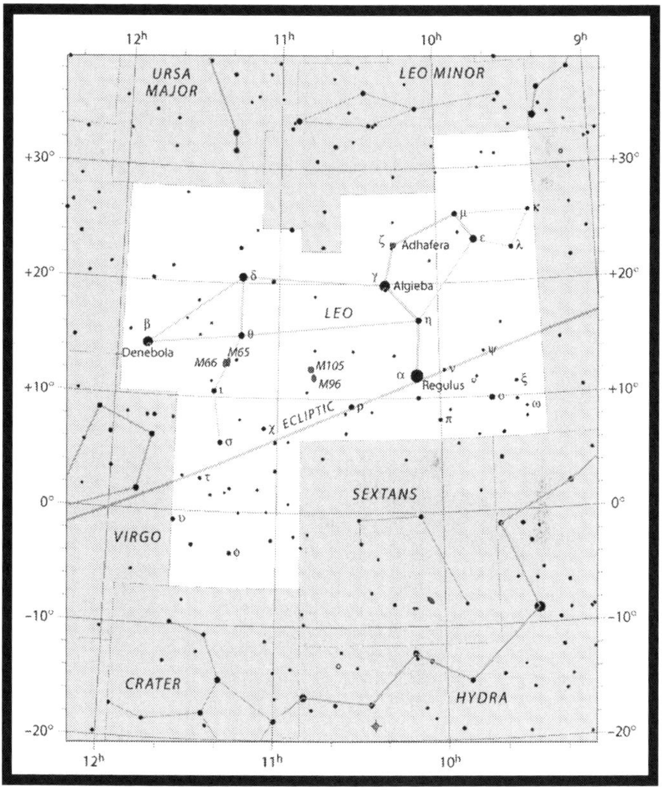

- Best observed in **Spring**
- There are **two** meteor showers associated with Leo
- The brightest star in the constellation is **Regulus**

COMA BERENICES

THE LION'S TAIL

Coma Berenices is a bit of a special constellation. It was mentioned by Ptolemy as an asterism, as he saw this pattern of stars as the tuft at the end of **Leo's** tail. However, another astronomer by the name of **Tycho Brahe** made it a constellation officially in the 16th century. It is the only one of the 88 current constellations to be named after a historical figure (Queen Berenice II of Egypt).
Coma Berenices is quite small, but contains a great amount of deep sky activity. It hosts a galactic supercluster, two galactic clusters, eight Messier objects and several star clusters.

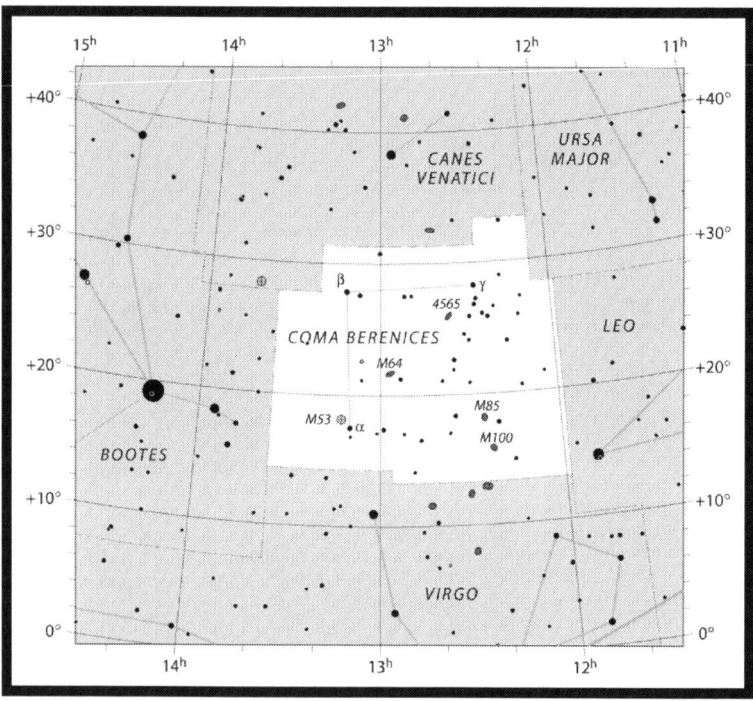

- Best observed in **Spring**
- The **Coma Berenicids** happen in Coma Berenices
- The brightest star in the constellation is **Beta Comae Berenices**

DRACO

THE DRAGON

Draco is a circumpolar constellation, which means it never sets under the horizon thus can be observed at any time of the year for observers in the northern hemisphere!
The Dragon is the 8th largest constellation in the night sky, and similarly to **Hydra**, it contains many deep sky objects due to its size.

The most famous objects located in Draco are the **Cat's Eye nebula** (NGC 654), the **Spindle galaxy** (M102) and the **Tadpole nebula** (Arp 188).

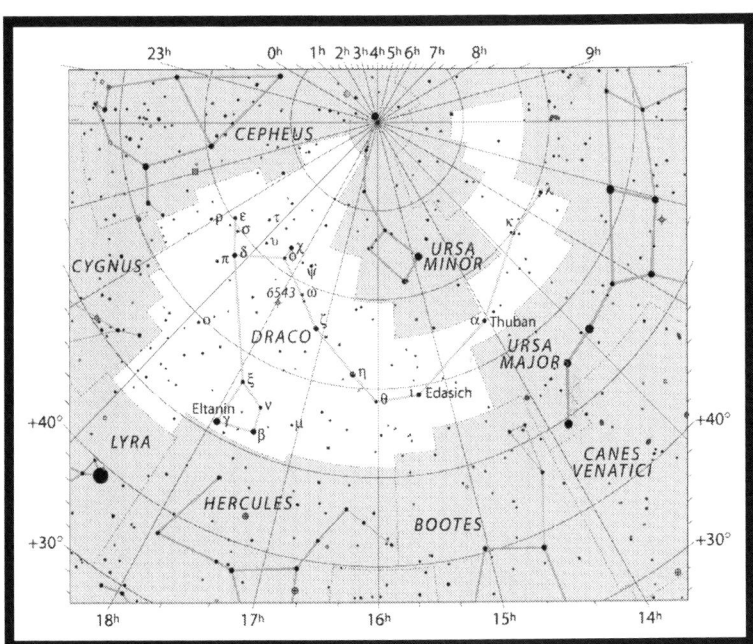

- Can be observed during **any** season of the year from the Northern hemisphere
- The **Draconids** happen in Draco
- The brightest star in the constellation is **Gamma Draconis**

THE LESSER AND GREATER BEAR

The Great Bear (**Ursa Major**) and the Lesser Bear (**Ursa Minor**) are two of the most famous constellations in the sky.

Their stories involve **Callisto**, a charming nymph who swore her chastity to the goddess Artemis. One day, Callisto met Zeus, and could not resist the temptation. The nymph broke her vow and became pregnant with the king of the gods.
She later gave birth to a boy, whom she named **Arcas**.

Infuriated, Zeus' wife Hera (who also tried to kill **Hercules** several times) transformed Callisto into a bear forever…

Almost 20 years passed since the day she was turned into an animal. Callisto had been living in the woods and fleeing from hunters, until she ran into someone familiar: her son.

Not recognizing his mother, Arcas panicked at the sight of the bear and aimed his spear towards the animal. Zeus, watching from Olympus, saw the tragedy about to happen and stopped Arcas by summoning powerful winds.
The gusts were so strong that both Callisto and Arcas flew up to the sky. Zeus decided to turn Callisto into Ursa Major, and her son into Ursa Minor*.

Hera, even more angry by her husband's actions, met with Oceanus and Tethys, and convinced them to never allow the bear to bathe in the northern waters. This is the reason why from perspectives in mid-northern latitudes, Ursa Major never sets below the horizon.

*In a different version of the myth, Arcas is turned into the constellation of **Boötes**. The page following Ursa Minor is Boötes, linked by the pursuit of the bears with his dogs, and is the last of Ptolemy's constellations mentioned in this section.

URSA MAJOR

THE GREATER BEAR

Ursa Major is a large constellation that is visible most of the year for observers in the northern hemisphere, and is one of the oldest known. The constellation of the Greater Bear is often one of the first ones taught to children, using its famous asterism, the "**Big Dipper**" (also called the "Plow").

Several interesting deep sky objects lie within Ursa Major, including seven Messier objects, such as **M81** (Bode's galaxy), **M82** (The Cigar galaxy), **M97** (The Owl nebula) and **M101** (The Pinwheel galaxy).

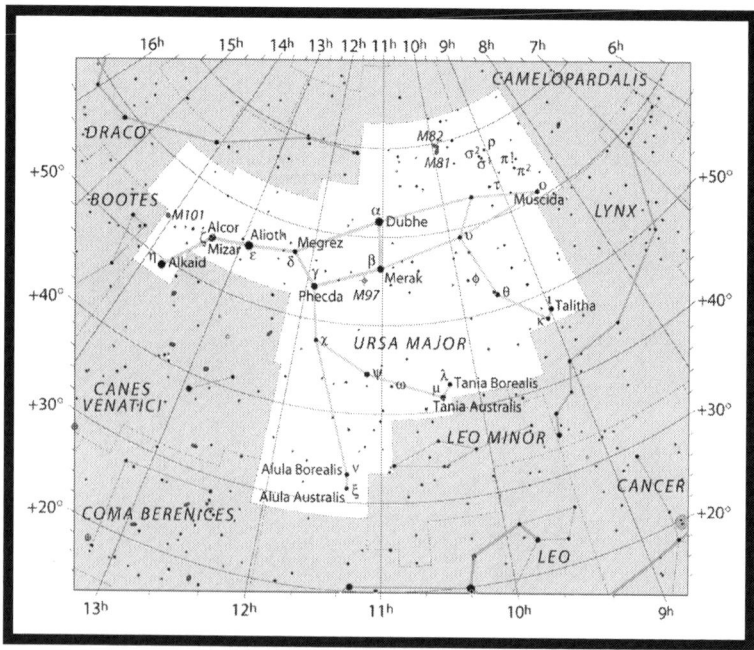

- Best observed in **Spring**
- There are **two** meteor showers associated with Ursa Major
- The brightest star in the constellation is **Alioth**

URSA MINOR

THE LESSER BEAR

Separated from **Ursa Major** by **Draco's** tail, Ursa Minor is very similar in shape to the Greater Bear, and also has its own asterism, known as the "**Little Dipper**". Ursa Minor is well known to astronomers living in the northern hemisphere. The star at the end of the bear's tail is the famous North Star, **Polaris**, which is used to polar align astronomy equipment. Ursa Minor is almost completely void of deep sky objects.

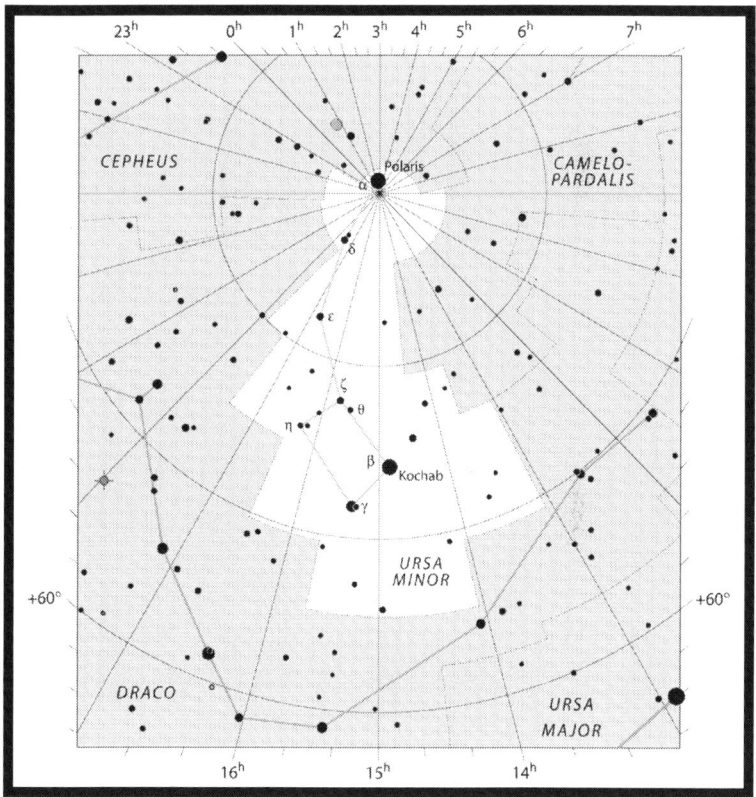

- Can be observed during **any** season of the year from the Northern hemisphere
- The **Ursids** happen in Ursa Minor
- The brightest star in the constellation is **Polaris**

BOÖTES

THE PLOWMAN

Home to one of the best meteor showers, the Quadrantids, and the fourth brightest star in the sky, Arcturus, is the constellation of Boötes. The only interesting target for amateur astronomers in the constellation is the **Great Void**.

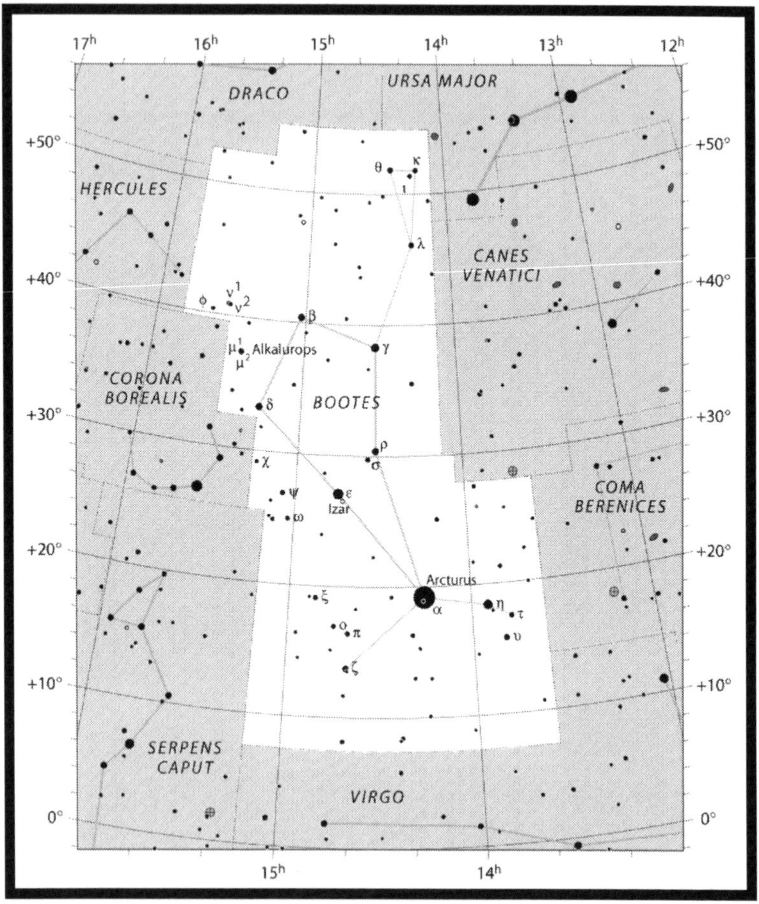

- Best observed in **Spring**
- There are **three** meteor showers associated with Boötes
- The brightest star in the constellation is **Arcturus**

FROM GREECE TO POLAND

It is with **Boötes**, the plowman in Greek mythology, that we end our journey on the 48 constellations introduced by the Greco-Roman astronomer, **Ptolemy**.

Taking a great leap from the 2nd to the 17th century, take heed and learn of the 7 constellations introduced by the Polish astronomer, **Johannes Hevelius**.

The logical transition from Ptolemy's mythology-themed constellations to Hevelius' is **Canes Venatici**.
Latin for "The Hunting Dogs", Canes Venatici is the only one out of the Polish patterns of stars that can be linked to Greek mythology.

Besides the dogs, Johannes Hevelius named four of his other constellations after **animals.** The last two, on the other hand, were named after… useful **objects**, which will also help in the logical transition to the third section of this book.

Since the Polish constellations can not be associated with Greek mythology, we are going to classify them by their location in the sky. There are only seven, so it will be a relatively short section!

Before discussing Hevelius' constellations, take a look at the following two pages for use as a guide to where you will be able to picture, at a glance, how the 48 constellations introduced by Ptolemy interact with one another.

VISUALIZING
PTOLEMY'S
48
CONSTELLATIONS

LEGEND

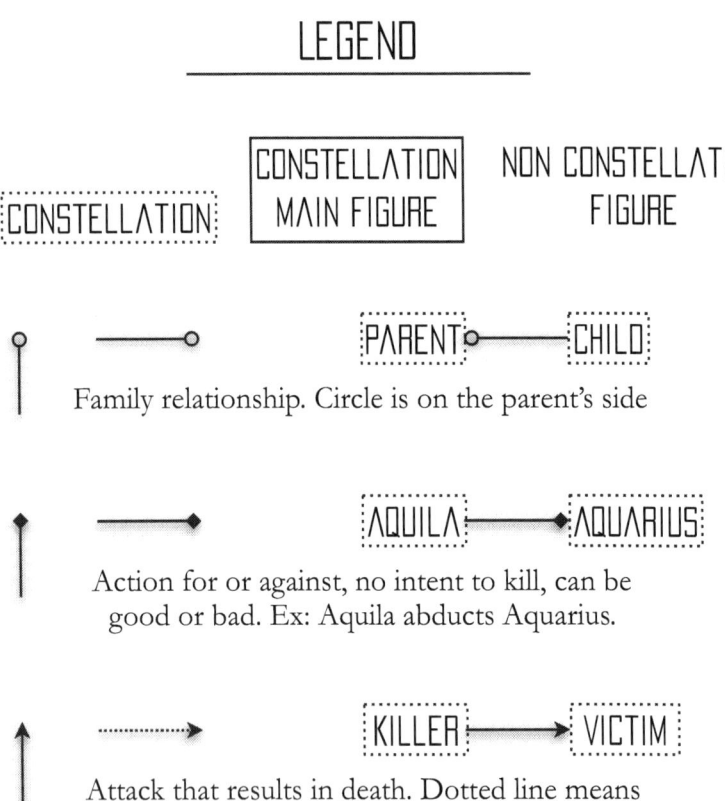

CONSTELLATION

CONSTELLATION MAIN FIGURE

NON CONSTELLATION FIGURE

PARENT ○——○ CHILD

Family relationship. Circle is on the parent's side

AQUILA ◆——◆ AQUARIUS

Action for or against, no intent to kill, can be good or bad. Ex: Aquila abducts Aquarius.

KILLER ——▶ VICTIM

Attack that results in death. Dotted line means the kill was unsuccessful or still in progress.

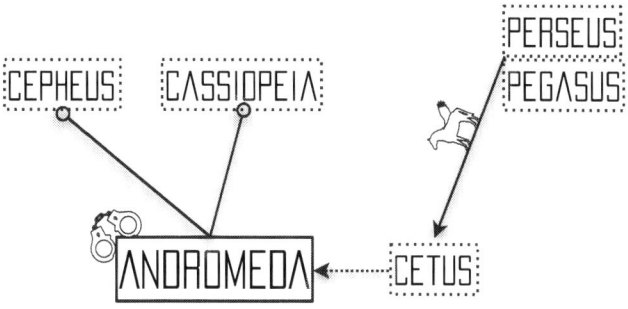

Andromeda, daughter of **Cepheus** and **Cassiopeia**.
Chained to a rock to be fed to **Cetus**, the sea monster.
Saved by **Perseus** and his flying horse **Pegasus**.

Transitioning to other water related
constellations with the attack of

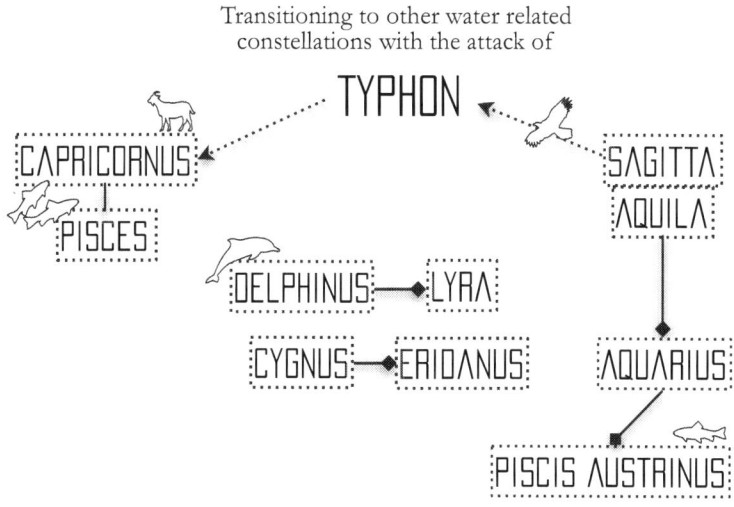

Capricornus and **Pisces** escape from Typhon.
Zeus kills Typhon with his arrow **Sagitta**, brought by **Aquila**.
Aquila abducts **Aquarius**, who now pours water on **Pisces**.
Delphinus saves Arion and his **Lyra** from the boat.
Cygnus recovers his friends's body in the **Eridanus** river.

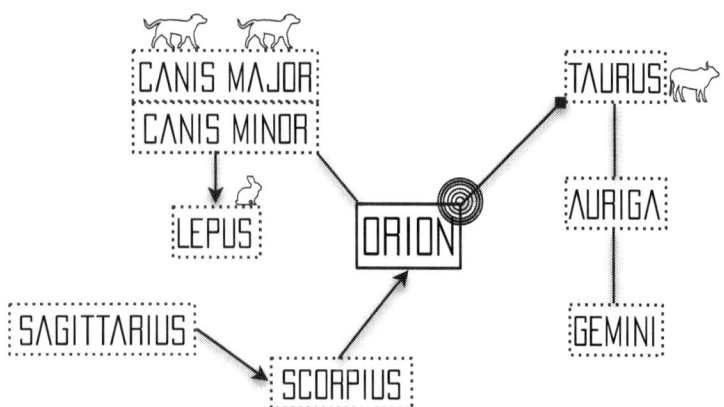

Orion, the Hunter, in love with the Pleiades in **Taurus**.
Auriga helped Orion regain his sight, close to **Gemini**.
Canis Major and **Minor**, dogs of Orion, chasing **Lepus**.
Sagittarius, aims a bow at **Scorpius** after it killed Orion.

Aries, the ram, and **Triangulum** are two constellations close to Pisces.

Ophiuchus is the serpent-bearer. **Serpens** is the only constellation to be in two different patches of sky.

Corona Borealis is the Northern Crown, made by Auriga and threw into the sky. **Corona Australis** is the Southern Crown.

Virgo represents the goddess of justice Dike, holding the scales **Libra**.

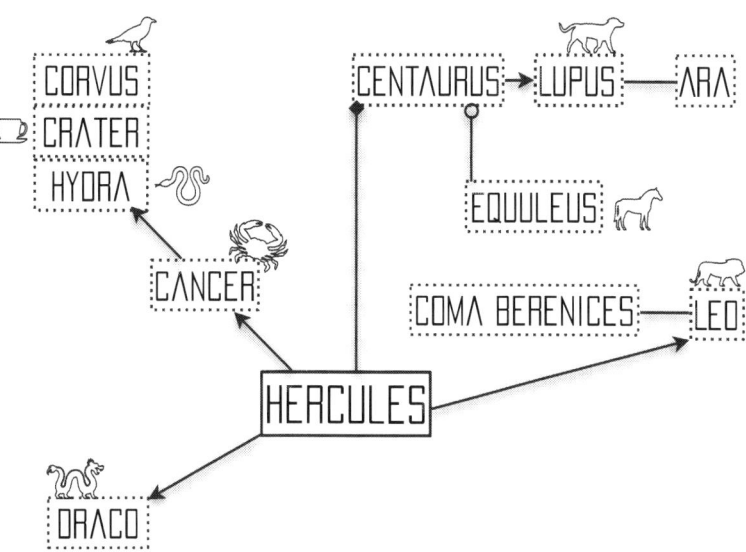

Corvus, the crow, blamed **Hydra** for being late with **Crater**.
Hercules kills **Cancer** and **Hydra** during one of his labors.
Hercules kills **Leo**, **Coma Berenices** is the lion's tuft.
For yet another task, Hercules kills **Draco**, the dragon.
Equuleus is hiding from her father **Centaurus**.
Centaurus sacrifices **Lupus**, the wolf at the altar, **Ara**.

Callisto, turned into a bear, **Ursa Major** by Hera after giving birth with Zeus' child.
Callisto's child, Arcas, transformed into **Ursa Minor** and reunited with his mother after almost 20 years.
Boötes, the plowman, chasing both bears with his dogs.

JOHANNES
HEVE

17th Century
Southern Hemisphere
5 Animals, 2 objects

LIUS

7
CONSTELLATIONS

CANES VENATICI

THE HUNTING DOGS

Canes Venatici is the next in the logical sequence of constellations in this guide, as it is said to be linked to **Boötes**, by representing his dogs chasing the Great Bear (**Ursa Major**).

Canes Venatici is a small constellation but full of activity! It is home to five Messier objects, including the very famous **M51** (Whirlpool Galaxy), a galaxy absorbing another.

In Canes Venatici you can also observe "**La Superba**", one of the reddest stars in the entire night sky. In 1988, astronomers discovered the largest void known, an area completely empty of galaxies, which they called "**The Great Void**" and also affects Boötes.

Boötes & his dogs: Canes Venatici

LOOKING UP

Canes Venatici is not easy to find in the sky. The best way to spot it is by first finding the Big Dipper asterism, and imagine the dogs right behind it. The bright star Cor Caroli is located where the bottom dog is.

- Best observed in **Spring**
- The **Canes Venaticids** peak every January
- The brightest star in the constellation is **Cor Caroli**

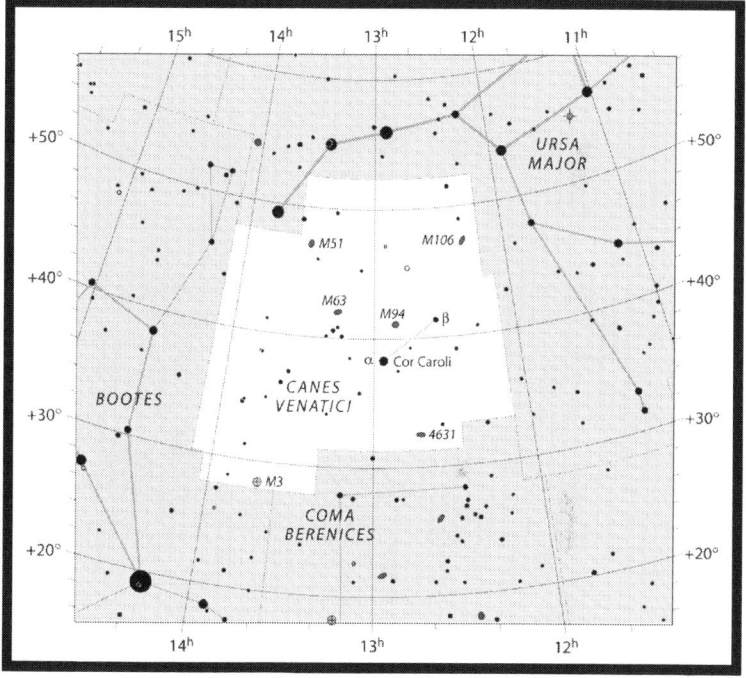

NOTABLE OBJECTS	M3	M51	M63	M106	NGC 4631
TYPE	CLUSTER	GALAXY	GALAXY	GALAXY	GALAXY
CATEGORY	GLOBULAR	SPIRAL	SPIRAL	SPIRAL	SPIRAL

LEO MINOR & LYNX

TWO FIERCE FELINES

Just above **Leo** is Leo Minor, a small and faint constellation that contains several visually interesting galaxies. The constellation is home to the Leo Minorids meteor shower, that peaks every October.

Leo Minor can be seen following another feline: Lynx. Johannes Hevelius gave it this name because of how faint its stars are. Johannes also challenged anyone to find it in the sky, and believed only those with the eyes of a Lynx would be able to point at it. **NGC 2419**, an intergalactic globular cluster that lies outside of our Milky Way galaxy, is visible in the constellation.

On the map below you can see where the two constellations are located compared to **Ursa Major**, **Leo** and **Gemini**.

Leo Minor (left) and Lynx (right)

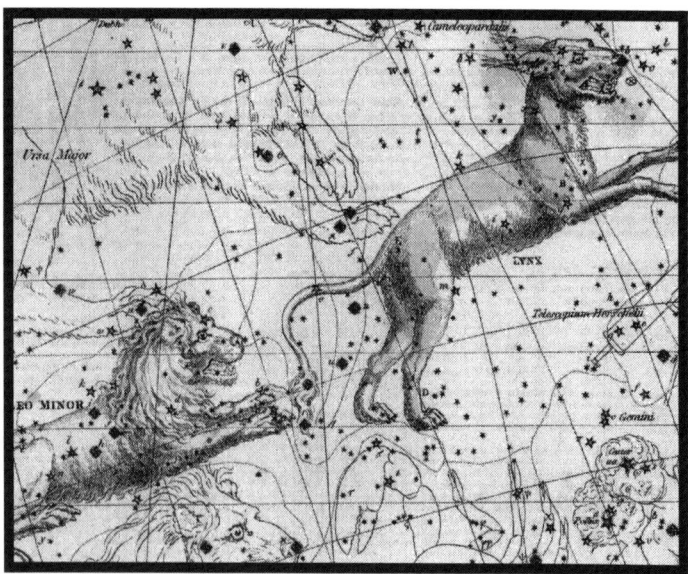

LACERTA

THE LIZARD

Lacerta, Latin for "Lizard" is located between **Andromeda** and **Cygnus**. It is small, faint and can be tricky to find. The easiest way to spot it is by its brightest stars. To the naked eye, it is in the shape of a W, or M, which is why Lacerta is also sometimes called "The Little **Cassiopeia**"!

NGC 7243 is the only interesting deep sky object in Lacerta. It is an open star cluster with a magnitude of 6.4, not far from the brightest star in the constellation.

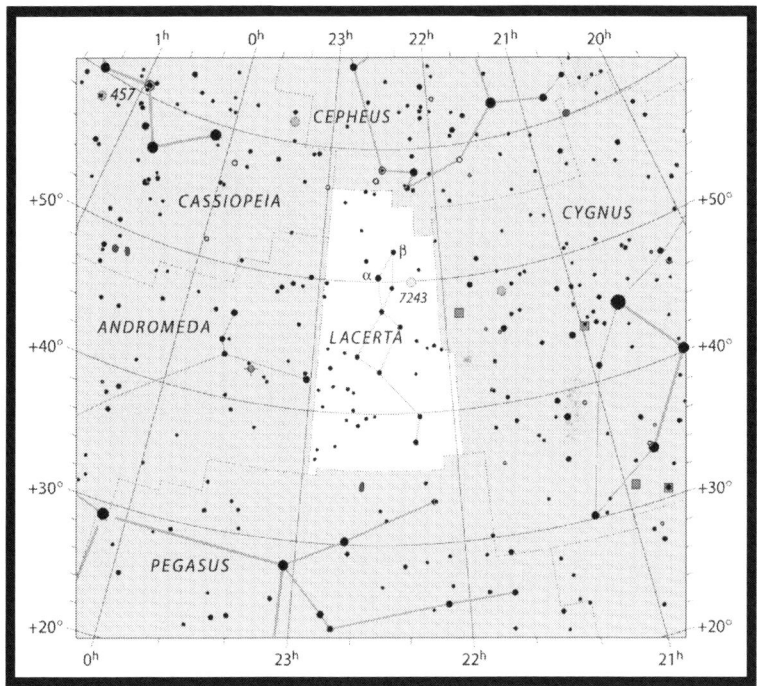

- Best observed in **Fall**
- There are **no** meteor showers associated with Lacerta
- The brightest star in the constellation is **Alpha Lacertae**

VULPECULA

THE FOX

Vulpecula is also a small and faint constellation not far from **Cygnus**.
Meaning "Little Fox" in Latin, Vulpecula lies inside the famous **Summer Triangle,** and is not very hard to find.

The constellation contains one very good target for amateur astronomers and astrophotographers: **Messier 27** (The Dumbell nebula). M27 has a magnitude of 7.5 and was the very first planetary nebula discovered in 1764.

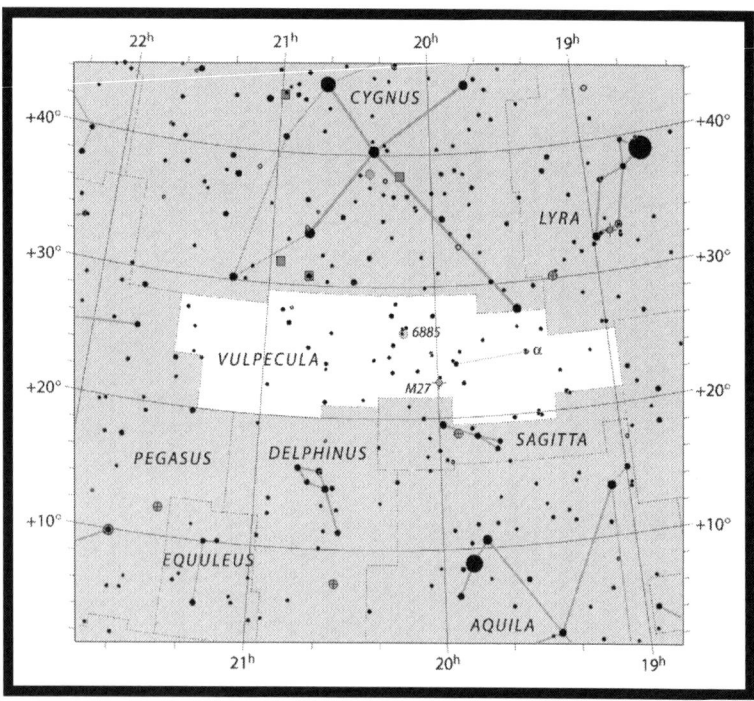

- Best observed in **Fall**
- There are **no** meteor showers associated with Vulpecula
- The brightest star in the constellation is **Anser**

SEXTANS

HEVELIUS' INSTRUMENT

Do you recall Hevelius' hatred for telescopes?

Sextans Uraniae, now known as just "Sextans", is a small and faint constellation named as tribute to the astronomical sextant instrument, which the Polish astronomer used to measure the different positions of the stars.
Sextans is located in a faint patch of sky, but is home to the famous **NGC 3115** (the Spindle galaxy), as well as other less impressive deep sky objects.

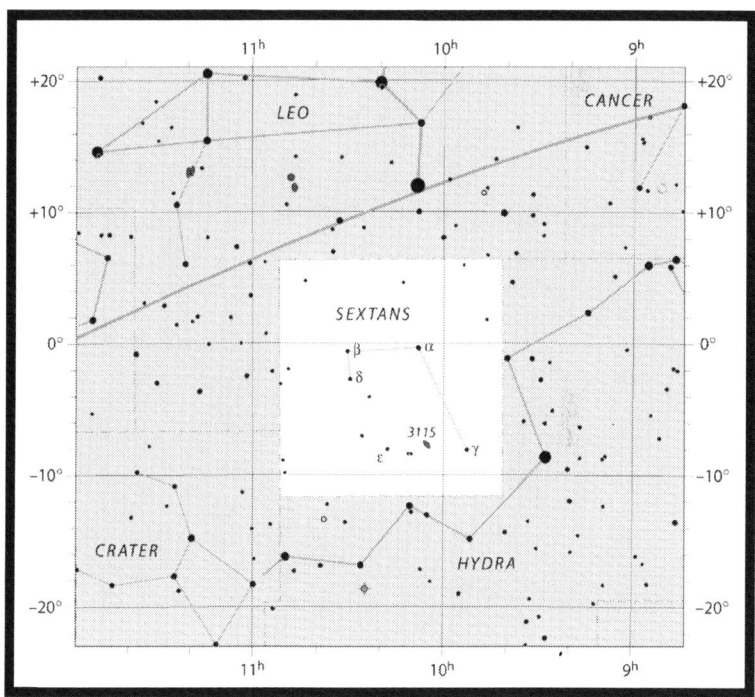

- Best observed in **Spring**
- The Sextantids meteor shower happens in Sextans
- The brightest star in the constellation is **Alpha Sextantis**

SCUTUM

SOBIESKI'S SHIELD

Originally called Scutum Sobiescianum, meaning "Shield of Sobieski", Scutum was introduced by Hevelius a year after Polish King Jan III Sobieski' victory in the Battle of Vienna in 1683.

Scutum is one of the smallest constellations in the night sky, but contains two famous Messier clusters, **M11** (The Wild Duck cluster) and **M26**.
Pioneer 11, the first probe to pass by Saturn, was launched by NASA on April 6, 1973, but is now out of power and drifting towards Scutum's stars.

LOOKING UP

Scutum is a faint constellation, with its brightest star, Alpha Scuti, shining at a magnitude of 3.85. It is located between **Aquila**, **Ophiuchus**, and **Sagittarius**.

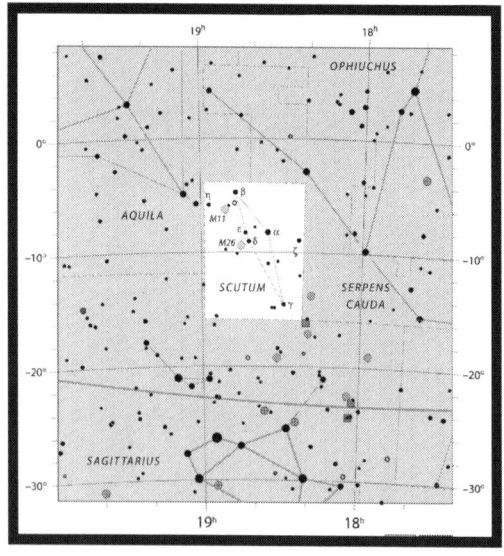

THE BATTLE OF VIENNA

Fought by King Jan III Sobieski's men against the Ottomans, the battle occurred on September 12, 1683. This was a decisive battle for not only Poland but several countries, as a Turkish win meant a rise of Islam all over Europe.
Scutum is the only remaining constellation to have been named after someone who was alive when it was introduced.

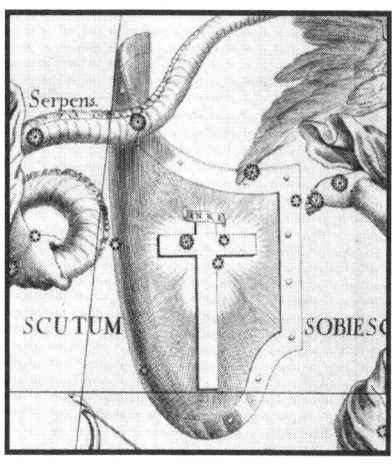

John III Sobieski and his shield during the Battle of Vienna

VISUALIZING HEVELIUS' 7 CONSTELLATIONS

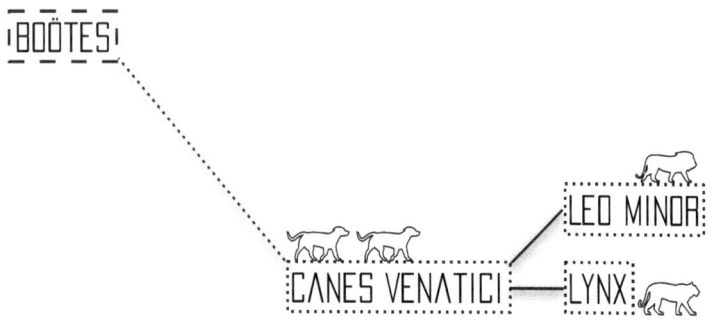

Canes Venatici, hunting dogs of **Boötes** (Ptolemy).
Leo Minor and **Lynx**, not far from Canes Venatici.

Lacerta (the lizard) and **Vulpecula** (the fox) are two small and faint constellations located on each side of Cygnus.

Sextans, Hevelius' favorite instrument, and **Scutum**, the shield, introduced in honor of his friend King Jan III Sobieski.

TO LACAILLE'S CONSTELLATIONS

Just as with **Ptolemy**, you will find a visual aid on the left page to memorize all of **Johannes Hevelius'** constellations. Although they are not as memorable as the constellations of the Greeks, it is not difficult to memorize them from the connections made. Not to mention there are 41 less than before!

Departing from the connections in Poland, the next section shifts to South Africa, more precisely to the Cape of Good Hope. This is where the French astronomer **Nicolas-Louis de Lacaille** landed when sent on a mission to record the stars.

Lacaille introduced a total of 14 new constellations, which he named after tools of arts and science used during his time, the **Enlightenment Era**. Most of those constellations are in honor of instruments he used, such as **Telescopium** (his telescope), **Octans** (his reflecting octant), **Horologium** (the pendulum clock he used for timing his observations), and more!

Lacaille is now credited for 17 constellations, as the astronomer decided to split one of the Greeks' constellations, **Argo Navis** (the ship) into three: **Carina** (the keel), **Puppis** (the poop deck), and **Vela** (the sails).

The 17 constellations are far from exciting as compared to the ones Ptolemy introduced, mainly because they are not as bright and most of them do not contain objects that northern observers are familiar with.

In order to create more memorable devices the next section will continue to link constellations with one another. The efforts into creating a constellation chain will be shown in the next pages! Now, let us turn our attention to…

NICOLAS-LOUIS DE
LACA

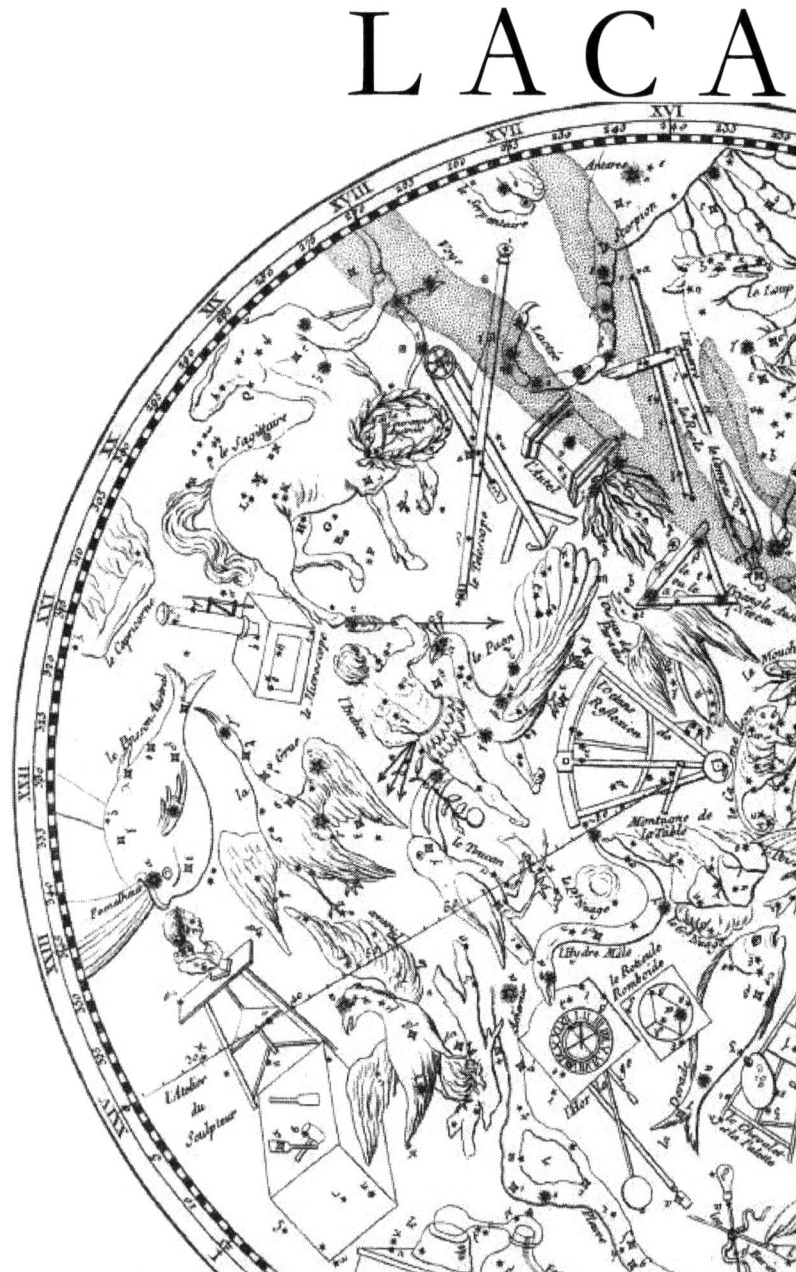

18th Century
Southern Hemisphere
Enlightenment Era Instruments

ILLE

17
CONSTELLATIONS

OCTANS

THE REFLECTING OCTANT

The name "Octans" stands for the reflective quadrant, the octant instrument, which Nicolas-Louis de Lacaille used almost every night to measure the position of the stars from the Cape of Good Hope.

Octans is the south equivalent of **Ursa Minor**, only much fainter and smaller. Octans contains the south pole star, **Sigma Octantis**, which means the constellation never sets below the horizon if observed from the southern hemisphere.

Octans does not contain any known deep sky objects.

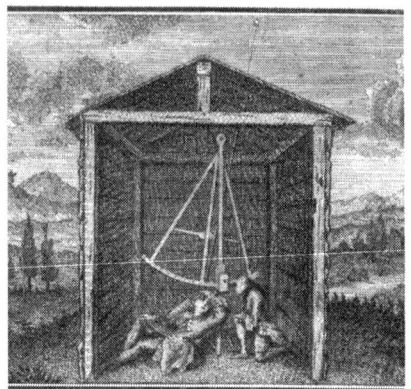

Lacaille using his reflecting octant

LOOKING UP

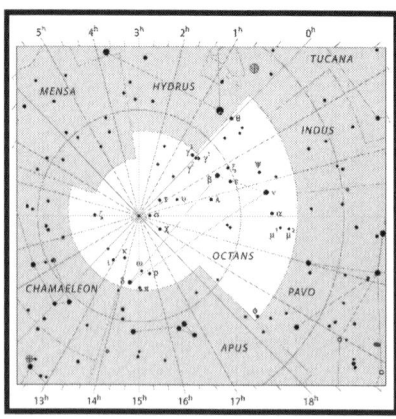

- Can be observed during **any** season of the year from the southern hemisphere
- There are **no** meteor showers associated with Octans
- The brightest star in the constellation is **Nu Octantis**

MENSA

TABLE MOUNTAIN

Just next to **Octans** is Mensa, another one of Lacaille's faint southern constellations. The only interesting object in Mensa, is the **Large Magellanic Cloud**, which gives the constellation a fuzzy look. Because of that, Lacaille gave it the Latin name of "Table" to represent Table Mountain, the place above the clouds from where the astronomer observed most of his nights.

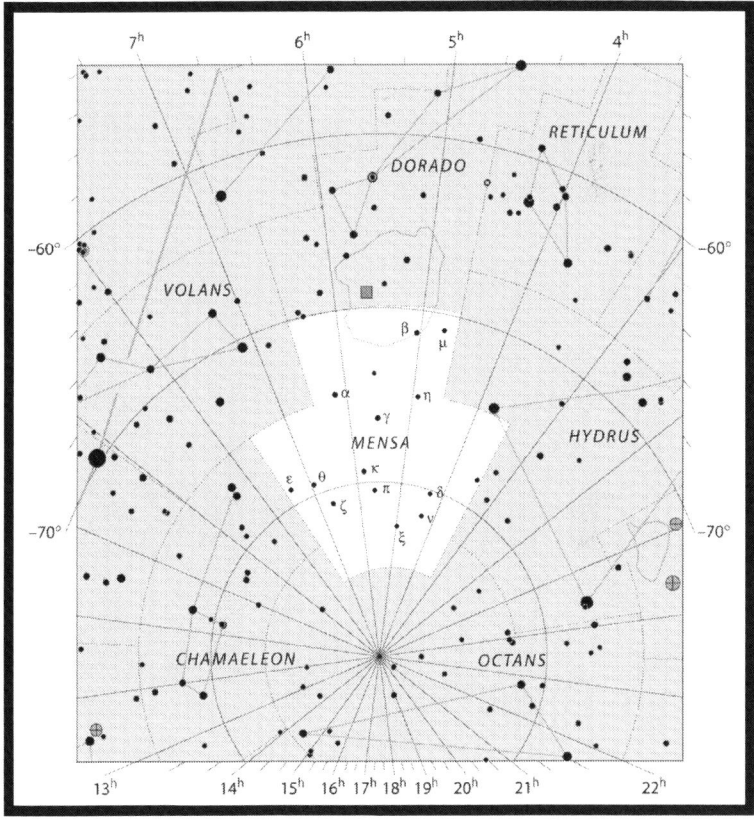

- Can be observed during **any** season of the year from the Southern hemisphere
- There are **no** meteor showers associated with Mensa
- The brightest star in the constellation is **Alpha Mensae**

HOROLOGIUM & RETICULUM

THE PENDULUM CLOCK AND THE TELESCOPE RETICULE

Horologium, meaning "Pendulum Clock" in Latin, was created by Lacaille to honor the clock he used to time his observations while observing from the Cape of Good Hope. Unlike most of his other constellations, this one is home to several galaxies and star clusters.

Introduced by Isaac Habrecht II under the name of Rhombus, Nicolas-Louis de Lacaille later renamed the constellation "Reticulum" which represents the reticle he had on his telescope. It is one of the smallest constellations of all, but contains some notable objects, such as the starburst **Topsy Turvy** galaxy.

The **Horologium-Reticulum Supercluster** is a huge galaxy supercluster spanning over a 12°x12° area of the sky. With about 5,000 galaxy groups, it contains more than 350,000 galaxies!

Horologium and Reticulum

TELESCOPIUM & MICROSCOPIUM

THE TELESCOPE AND THE MICROSCOPE

Telescopium (the telescope), and Microscopium (the microscope) are two constellations on each side of **Sagittarius**' heels.

Faint, small, and without any popular deep sky objects, they are two of the least interesting constellations in the entire night sky.

The stars forming Microscopium almost cannot be seen at all without a telescope or binoculars. Patrick Moore, a famous English astronomer, said that the constellation was "totally unremarkable".

Neither Telescopium or Microscopium is home to any meteor showers.

Microscopium (top left) and Telescopium (bottom)

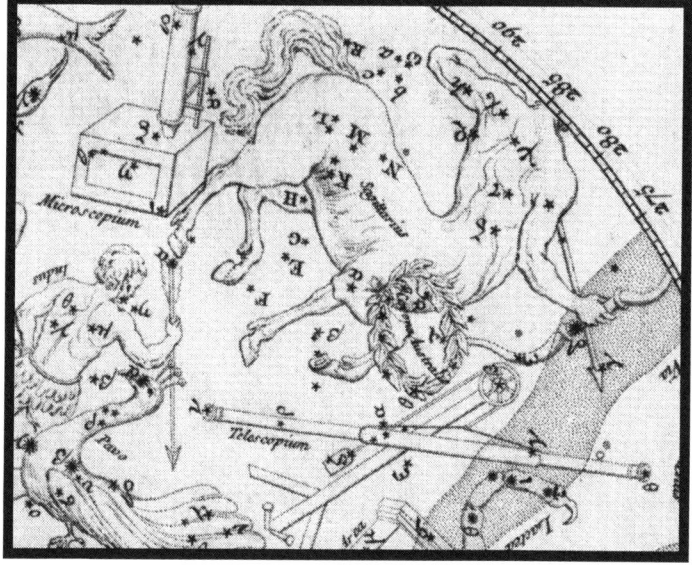

CAELUM & SCULPTOR

THE SCULPTOR'S CHISEL AND STUDIO

Next to **Horologium** is Caelum, or the Sculptor's Chisel. It is a very faint constellation and the eighth smallest in the sky. It does not contain notable deep sky objects and there are no meteor showers associated with it. The brightest star in the constellation is Alpha Caeli with a magnitude of 3.39.

On the other side of the river Eridanus, is another faint and small constellation… Sculptor. It represents the sculptor's studio and is also not home to any meteor showers. Its brightest star is Alpha Sculptoris at magnitude 4.30.

Unlike Caelum and most of Lacaille's constellations, Sculptor contains several visually impressive deep sky objects.
The lenticular and ring **"Cartwheel"** galaxy, although not very bright, looks stunning when photographed. It currently looks like the aftermath of its merging with another galaxy from about 300 million years ago.
In Sculptor you will find the southern **Cigar galaxy**, the **Silver Coin galaxy,** and the **Sculptor group** of galaxies!

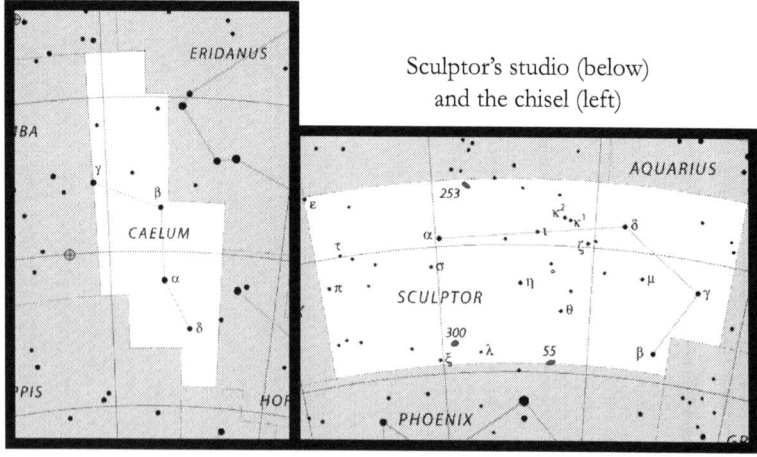

Sculptor's studio (below) and the chisel (left)

PICTOR

THE PAINTER'S EASEL

Next to Caelum is Pictor. Originally named the Painter's Easel (Equuleus Pictoris), the constellation was later shortened to Pictor.
It is located near the **Large Magellanic Cloud**. It is small, faint and a bit outshone by Canopus, the second brightest star in the night sky after Sirius and the brightest star in the southern hemisphere. Canopus is part of the Carina constellation.

The only interesting object for observers is **NGC 1705**, a dwarf peculiar lenticular galaxy with a white and blue hue.

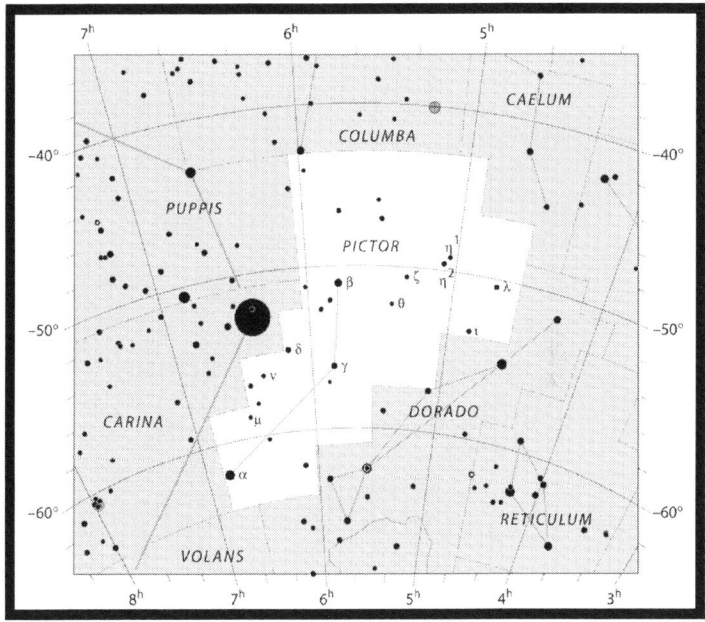

- Best observed in **Winter**
- There are **no** meteor showers associated with Pictor
- The brightest star in the constellation is **Alpha Pictoris**

FORNAX

THE FURNACE

Also near **Caelum** is Fornax, the furnace. It was first named Fornax Chemica to represent the small heater widely used in chemistry during Lacaille's time. Six of the discovered star systems in this medium-size constellation were confirmed to have exoplanets.

Fornax is home to the **Fornax cluster**, one of the richest cluster of galaxies "close" to ours, at 62 million light years away. The constellation also contains several more deep sky objects, mostly galaxies and star clusters.

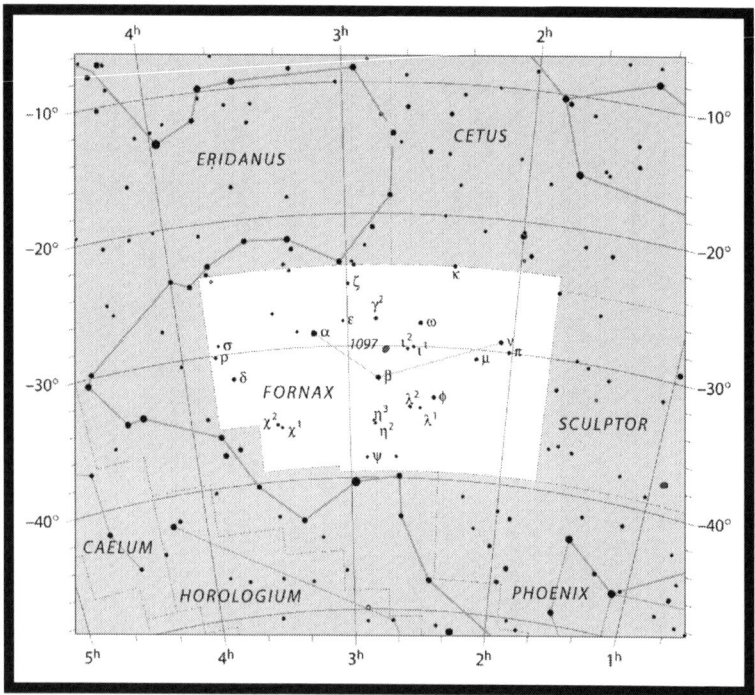

- Best observed in **Fall**
- There are **no** meteor showers associated with Fornax
- The brightest star in the constellation is **Alpha Fornacis**

ANTLIA

THE AIR PUMP

Antlia Pneumatica, now shortened to "Antlia", is a constellation near Centaurus and represents the air pump.

The stars in Antlia are very faint but its brightest star, Alpha Antliae, shines at a magnitude of about 4.25. Although faint as well, the constellation's deep sky objects are numerous. Antlia is home to the **Antlia Dwarf** galaxy, which wasn't discovered until 1997! There is also the **Antlia cluster of galaxies**, which is the third closest to Earth after the Virgo and Fornax clusters. Several other beautiful but faint galaxies can be found in Antlia.

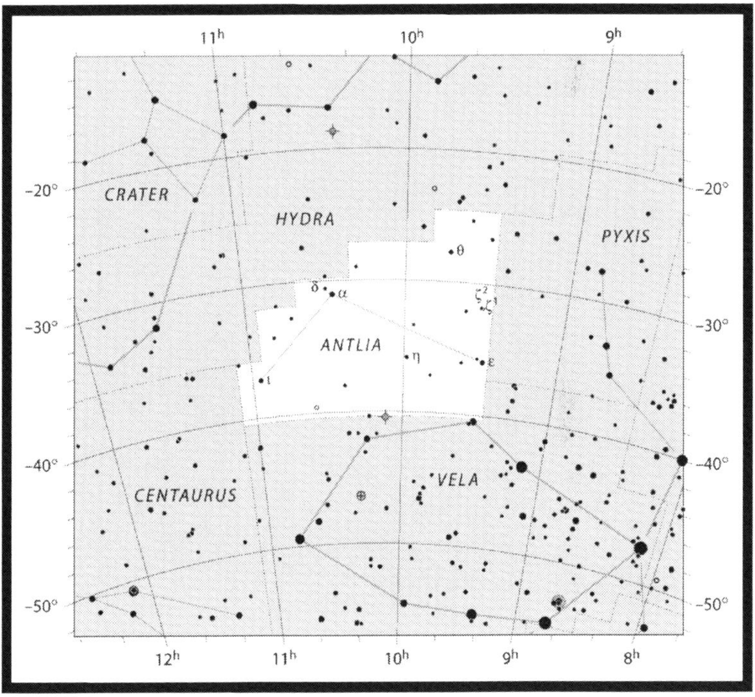

- Best observed in **Spring**
- There are **no** meteors shower associated with Antlia
- The brightest star in the constellation is **Alpha Antliae**

NORMA & CIRCINUS

THE SQUARE AND COMPASS

Norma and Circinus are two constellations really close to each other located between **Ara** and **Lupus**. Nicolas-Louis de Lacaille introduced these constellations to fill the void between the wolf and the altar.

Their name, the square and ruler (Norma), and the drawing compass (Circinus) were given due to the third constellation in their vicinity, **Triangulum Australe**.

Triangulum Australe represents another one of the draughtsman's instruments, the level. The constellation was drawn by Dutch explorers before Lacaille's time.

Norma and Circinus both contain a few interesting clusters and nebulae. The most famous one being the **Ant nebula**, a beautiful bipolar planetary nebula in the constellation of the set square.

Norma (left), Circinus (top right).
Part of Triangulum Australe can be seen on the right

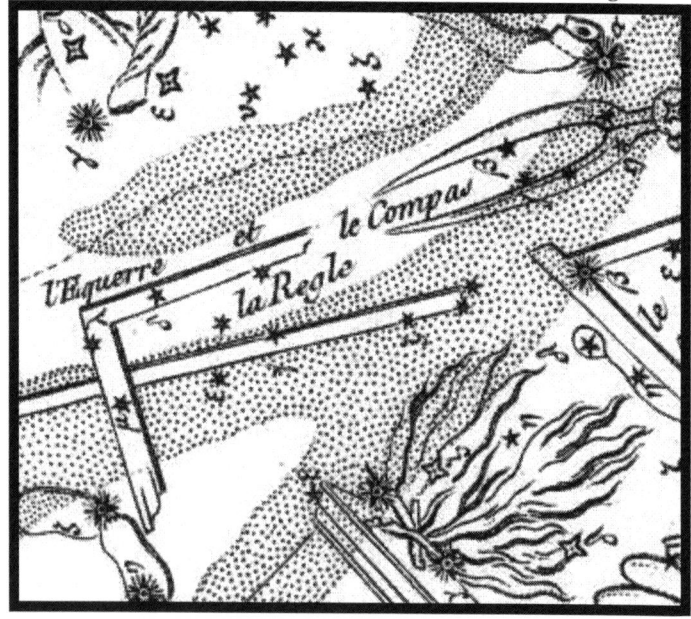

PYXIS

THE MARINER'S COMPASS

Pyxis is a small constellation but contains a few interesting objects, such as the planetary nebula **NGC 2818**, the barred spiral galaxy **NGC 2613**, and the open cluster of stars **NGC 2627**. Just like most southern constellations, Pyxis is not home to any Messier objects.

Pyxis is latin for "Mariner's compass" and is located just between **Puppis**, **Carina**, and **Vela**.
John Herschel proposed to rename Pyxis to "Malus", meaning the mast of the ship which is formed by Puppis, Carina and Vela. His idea was rejected by the Astronomical Union.

Location of Pyxis - Urania's Mirror

LOOKING UP

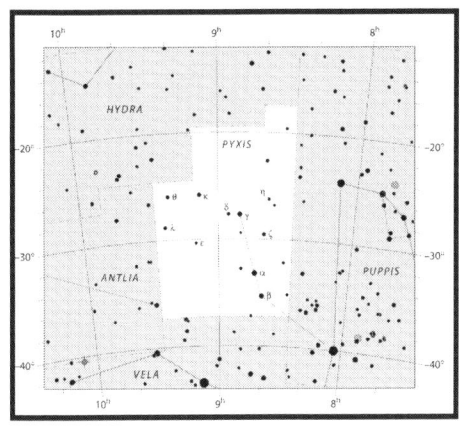

- Best observed in **Winter**
- There are **no** meteor showers associated with Pyxis
- The brightest star in the constellation is **Alpha Pyxidis**

105

THE SHIP OF THE HEROES

One of Ptolemy's original constellations was **Argo Navis**, the boat.

Argo Navis was named after Argus, who built the ship with the assistance of Athena. The goddess then added magical oak to the boat, which gave it the ability to speak.

When it finally was able to sail, Argo Navis was without a doubt the strongest ship ever built. On board were 50 of the greatest Greek heroes, such as Herakles (**Hercules**), Castor and Polydeuces (**Gemini**), and **Cepheus**.
Jason, the leader of the group, took the crew on a mission to retrieve the Golden Fleece. Passing through storms, rocks, violent waves and enemies, the ship made it to its destination: the island of Colchis. No other boat had ever been able to make the journey in the past.

The heroes retrieved the Golden Fleece without much trouble, and sailed off quickly, as they were being pursued by the Colchian army.

Due to the success of the mission, the ship was placed into the sky forever, as the large constellation of Argo Navis.

Well, not really forever…

Hundreds of years later, Nicolas-Louis de Lacaille, finding the huge and uneven shape of the constellation unappealing, took it upon himself to divide Argo Navis into three. **Carina** (the keel), **Puppis** (the poop deck) and **Vela** (the sails) were adopted by the Astronomical Union to essentially replace the old constellation. Lacaille also added **Pyxis** (the mariner's compass) later, where the mast of the boat is located.

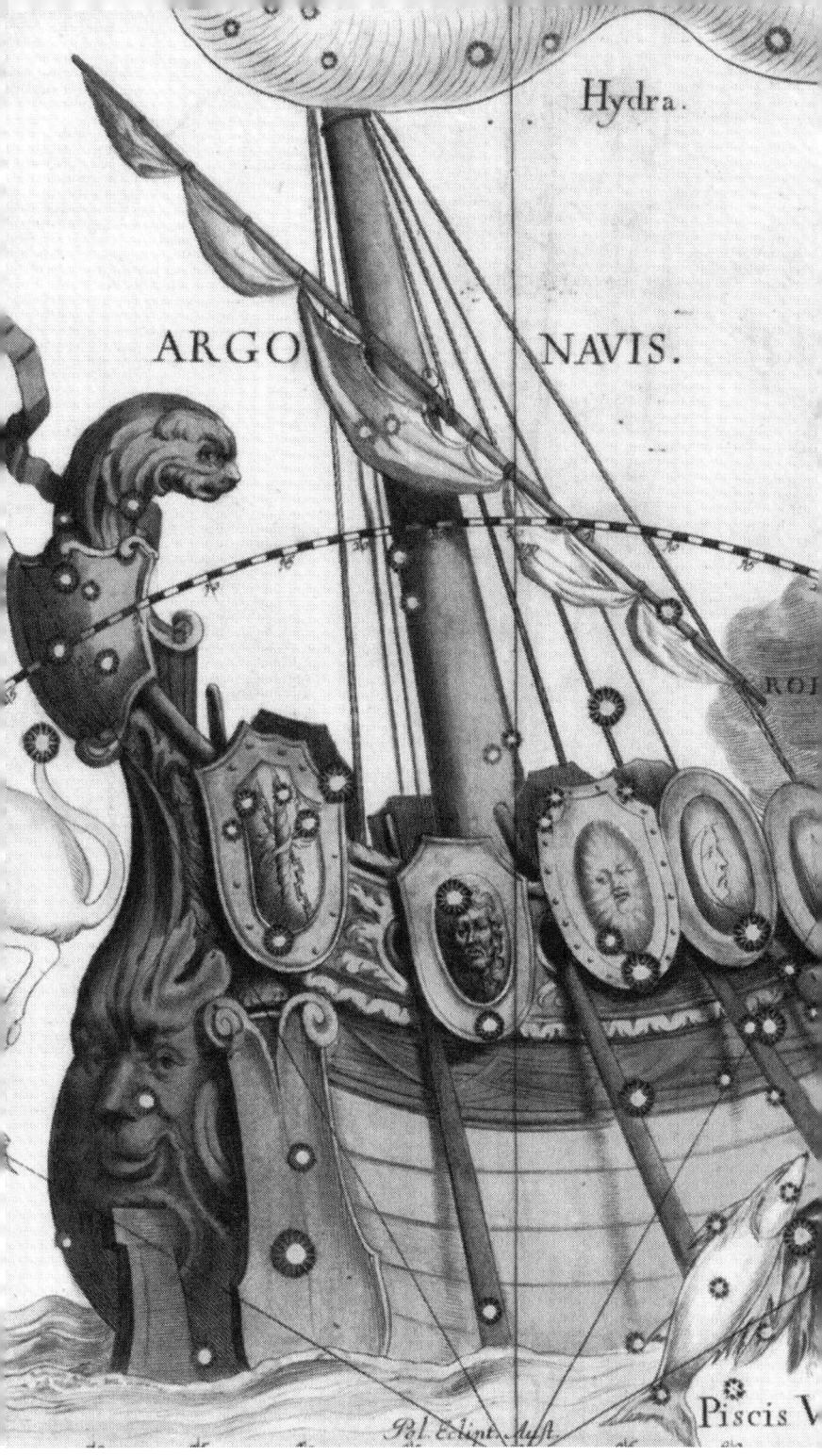

CARINA

THE KEEL

Carina is one of the most famous constellations of the southern sky. Its name is Latin for the Keel, and represents the base of the former Argo Navis ship.
Its brightest star is **Canopus**, a yellow/white supergiant that shines at a magnitude of -0.72, making it the second brightest star in the entire night sky, after Sirius, the Dog Star.
Carina is home to the **Eta Carinids** meteor shower, which peaks each year on January 21.

The constellation contains several visually impressive deep sky objects. Those include the very famous **Carina nebula** (NGC 3372), the **Wishing Well** cluster (NGC 3532), the **Diamond** cluster (NGC 2516), and more!
The Wishing Well cluster was the first object discovered by the Hubble Space Telescope in 1990. The two others listed above were both discovered by Lacaille himself in 1751-1752.

THE CARINA NEBULA

The Carina Nebula is a famous object in the southern hemisphere. It is bright, large, and is located in the constellation of Carina. Four times larger than the **Orion nebula**, NGC 3372 is a popular target for astrophotographers and observers. The Carina nebula is home to **Eta Carinae**, a hypergiant star with a luminosity of about four million times that of our Sun. You may also be able to spot **Trumpler 14**, one of the youngest star clusters known.

"Mystic Mountains" by NASA/Hubble

LOOKING UP

Carina is best observed in **Winter**, and can be found in between Volans, Vela, Musca, Puppis and Pictor.

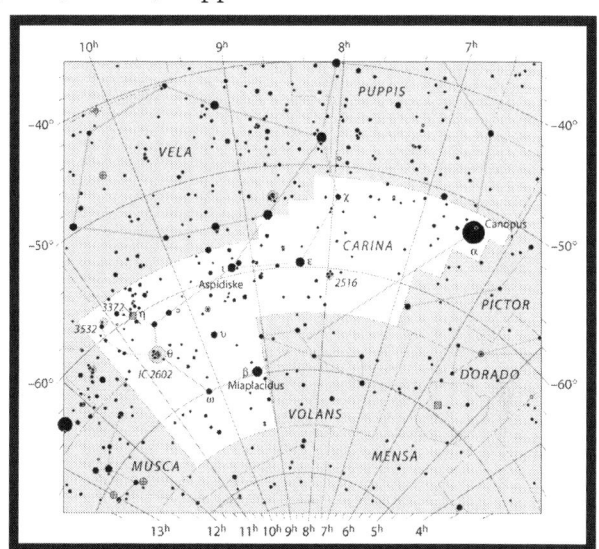

Argo Navis divided - Urania's Mirror

PUPPIS

THE POOP DECK

In between the two brightest stars of all, **Sirius** and **Canopus**, is the constellation of Puppis.
Puppis never rises high and is difficult to observe from northern locations. It contains several objects, including three open clusters from the Messier catalogue (**M46**, **M47**, and **M93**).

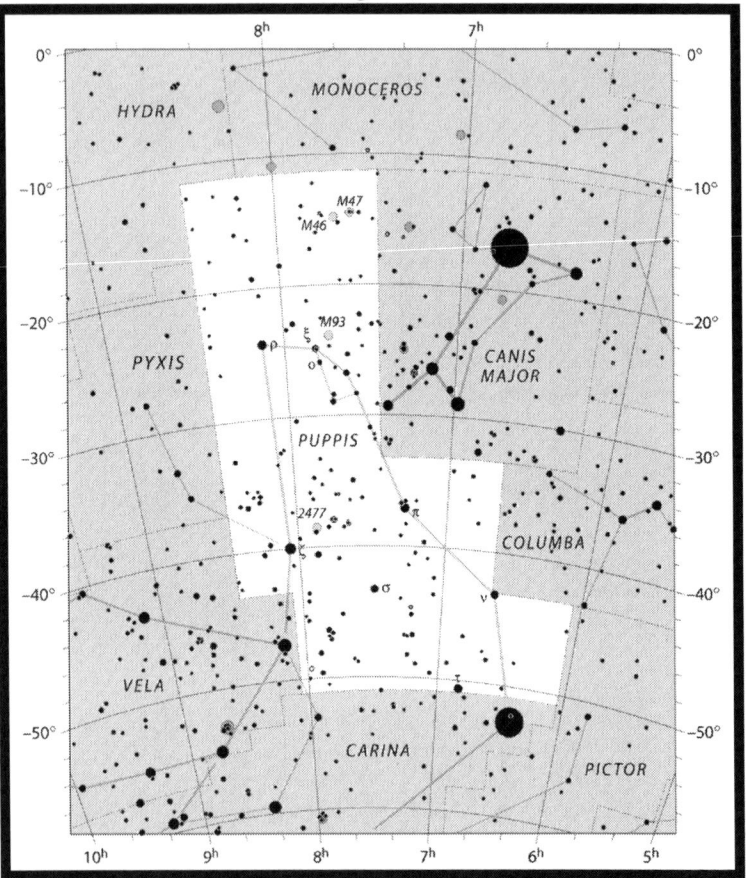

- Best observed in **Winter**
- There are **three** meteor showers associated with Puppis
- The brightest star in the constellation is **Naos, Zeta Puppis**

VELA

THE SAILS

Next to Puppis, Vela is another constellation that is not visible for northern observers, and does not contain any objects that Charles Messier could catalog.

You may be able to spot the "**False Cross**", an asterism in the constellation that is often mistaken as being the southern cross. There are many beautiful nebulae in Vela, such as the **Southern Ring** nebula (NGC 3132) or the **Gum nebula**. There are a few star clusters, the most impressive one being **NGC 2547**, discovered by Lacaille during his trip.

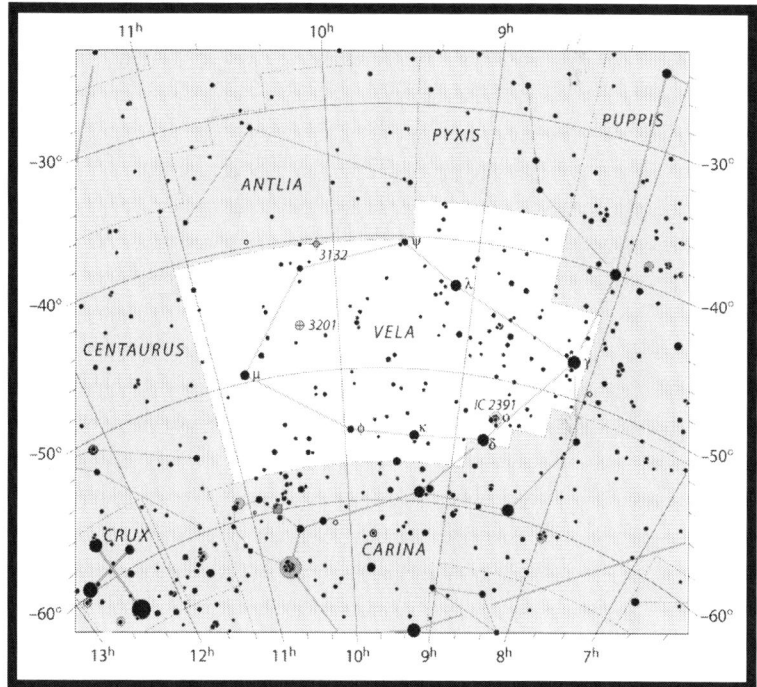

- Best observed in **Spring**
- There are **three** meteor showers associated with Vela
- The brightest star in the constellation is **Gamma Velorum**

`SEXTANS`
`SCUTUM`

Octans was one of Lacaille's main instruments during his time at the Cape of Good Hope.

`OCTANS`

`MENSA`

Mensa represents Table Mountain, one of Lacaille's observing spots above the clouds

Lacaille introduced **Horologium** in honor of another one of his instruments, the pendulum clock.

`HOROLOGIUM`

Next to it is **Reticulum**, representing the reticule that the astronomer installed on his telescope to measure data more precisely.

`RETICULUM`

Telescopium, representing the astronomer's most valuable instrument for deep sky observation. **Microscopium** represents, you guessed it… the microscope.

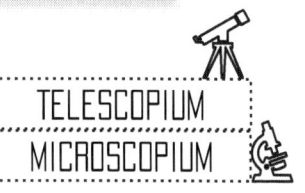

The Sculptor's chisel (**Caelum**) and his studio (**Sculptor**).

Next to the Sculptor's chisel is another constellation related to arts, **Pictor**, the painter.

Fornax was named after the invention of the furnace, **Antlia** after the air pump.

Norma, the square set and the ruler, **Circinus**, the drawing compass. Those two tools are right next to Triangulum Australe, introduced by Dutch astronomers.

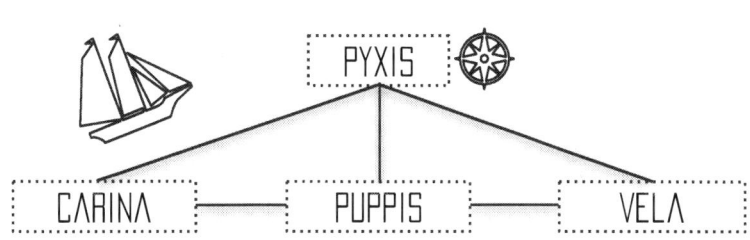

Carina, **Puppis**, and **Vela** all came from the big constellation of Argo Navis, the ship. They represent the Keel, the Deck, and the Sails. **Pyxis** was added where the mast is and stands for the Mariner's Compass.

Now, to the last destination, Denmark. Lastly, we meet the 16 constellations of **Petrus Plancius**, introduced after the voyage of **Keyser** and **de Houtman** in the East Indies.

PETRUS PLAN

With the help of

Pieter Keyser & **Frederick de Houtman**

16-17th Century Southern Hemisphere Indies Encounters

CIUS

16
CONSTELLATIONS

TRIANGULUM AUSTRALE

THE LEVEL

Triangulum Australe, or the "Southern Triangle", is similar in shape to the north **Triangulum** constellation only smaller. Triangulum Australe does not contain many deep sky objects, the most interesting one would be **ESO 69-6**, two beautiful merging galaxies captured by Hubble.

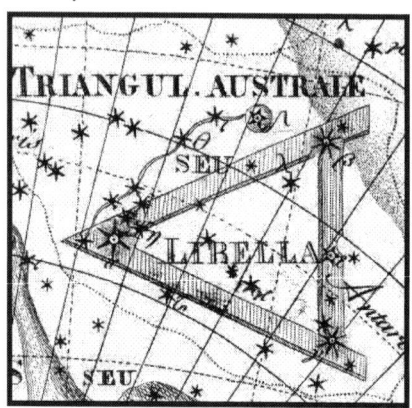

The constellation lies in the same area as **Norma** (the square set) and **Circinus** (the drawing compass). The Dutch astronomers did not really see anything that could represent the constellation. On the other hand, Nicolas-Louis de Lacaille noted it would make sense that it was another tool, so it is also known as the level.

Drawing of Triangulum Australe

LOOKING UP

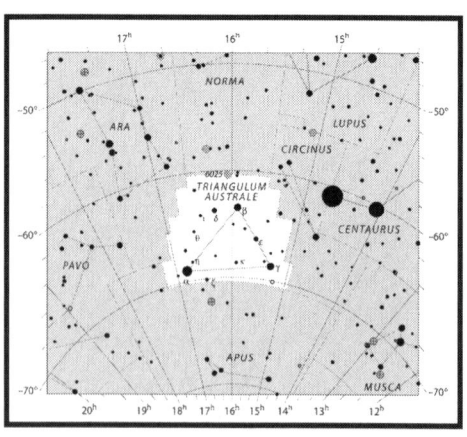

- Can be observed during **any** season of the year from the Southern hemisphere
- There are **no** meteor showers associated with Triangulum Australe
- The brightest star is **Atria**

CRUX

THE SOUTHERN CROSS

Having been part of **Centaurus** in the past, Crux is now its own constellation and is the smallest of all 88 in the entire night sky. Petrus Plancius is credited for dividing it from Centaurus as he was the first to draw the two distinctively on his globe.

The stars in Crux are very bright and form the iconic, easily recognizable shape of a cross. It lies in a busy area of the Milky Way band, and contains a couple of deep sky objects, the **Coalsack** nebula and the **Jewel Box** cluster of stars.

Crux is often associated with the cruxifixction of Christ.

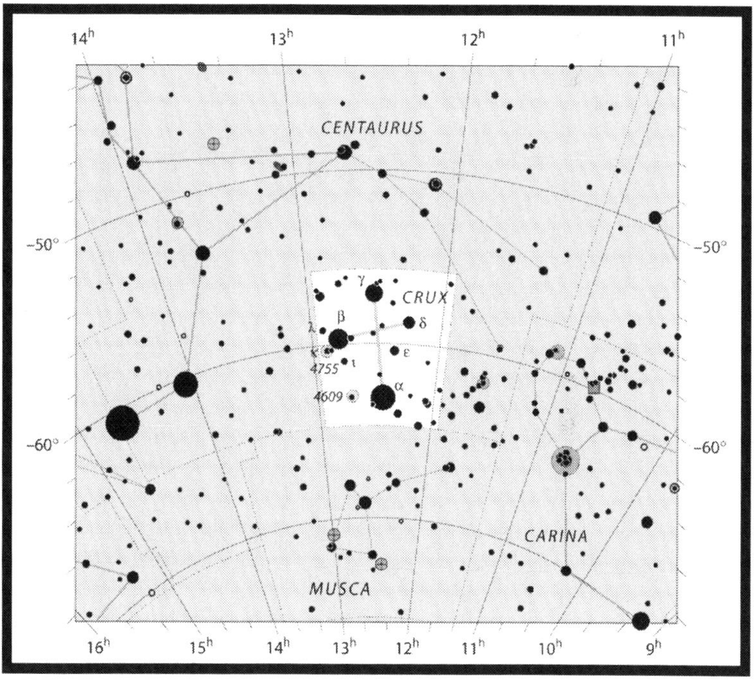

- Can be observed during **any** season of the year from the southern hemisphere
- The **Crucids** meteor shower happens in Crux
- The brightest star in the constellation is **Acrux**

COLUMBA

NOAH'S DOVE

Three constellations were introduced before the Dutch's voyage to the Indies: **Triangulum Australe**, **Crux** and **Columba**. Plancius really liked Argo Navis (the ship later divided by Lacaille into 3), and made Columba using some of the boat's stars. On his globe of the night sky, he renamed Argo Navis to "Noah's Ark" and stated that Columba represented Noah's dove, that came to the Ark with an olive branch in its beak, meaning the flood was ceasing and life was returning. There are not many bright deep sky objects in Columba, besides the globular cluster **NGC 1851**. The rest are faint galaxies.

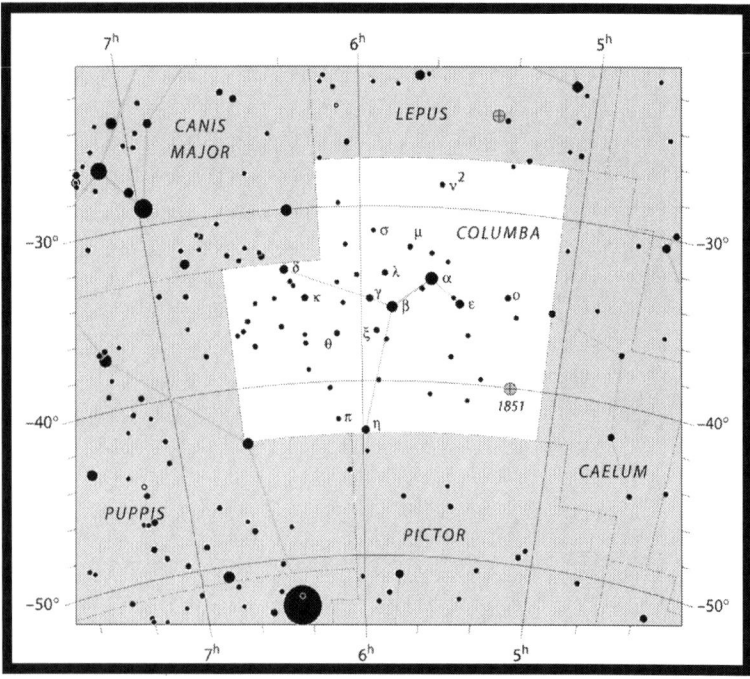

- Best observed in **Winter**
- There are **no** meteor showers associated with Columba
- The brightest star in the constellation is **Phact**

INDUS

THE NATIVE

Indus is a faint constellation located between **Octans** and **Microscopium**. Indus does not contain any bright deep sky objects, although some of its galaxies may be interesting to photograph with the right equipment.

Indus was named after the indigenous people the Dutch explorers would encounter while traveling in the East Indies.
The constellation was drawn first as a native man, holding arrows in each hand. He can be seen in a hunting position, aiming one of the arrows towards another one of Plancius' constellations, **Pavo** (the peacock).

Drawing of Indus

LOOKING UP

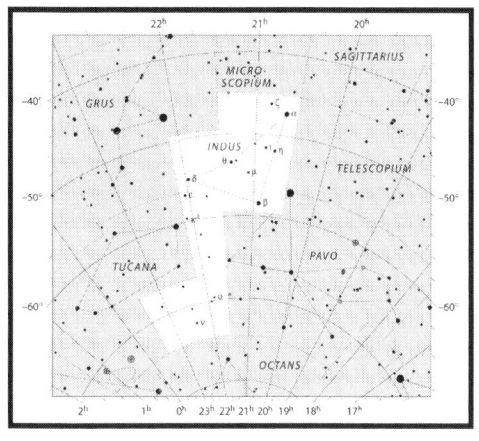

- Best observed in **Summer**
- There are **no** meteor showers associated with Indus
- The brightest star in the constellation is **Alpha Indi**

MUSCA

THE FLY

Musca is one of the smaller constellations listed (the 77th in size) and is not easily recognizable. However, it contains several interesting deep sky objects, such as the **Coalsack** nebula (which extends to **Crux**), the **Hourglass** nebula, the globular cluster **NGC 4372**, and others.
Petrus Plancius drew the constellation on his globe in 1598, but did not label it. Johann Bayer later named it Apis (meaning "the bee").
Several years passed until Nicolas-Louis de Lacaille decided to change the constellation's name to "Musca", meaning "the fly".

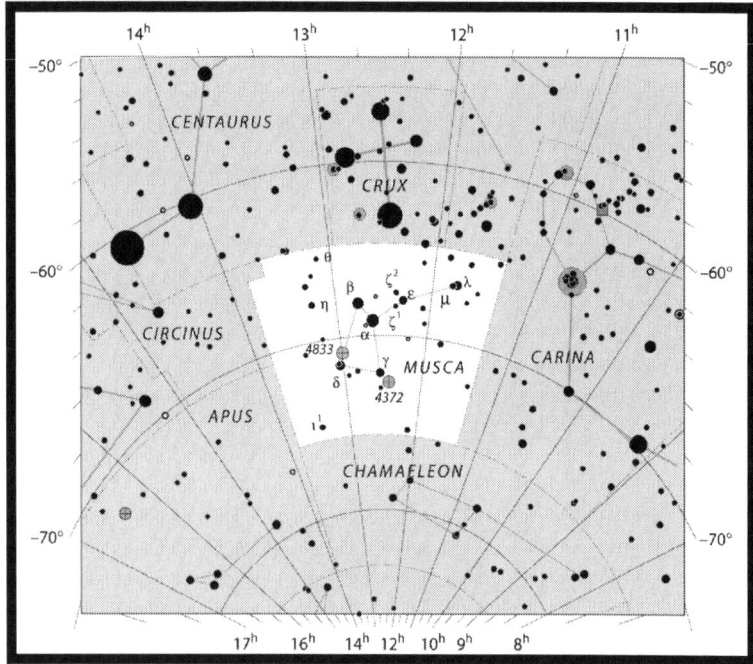

- Can be observed during **any** season of the year from the southern hemisphere
- There are **no** meteor showers associated with Musca
- The brightest star in the constellation is **Alpha Muscae**

CHAMAELEON

THE CHAMELEON

Chamaeleon, named after the lizard that can change colors, is located next to **Musca**, and represents the chameleon grabbing the fly with its tongue. The **Chamaeleon Cloud Complex** is a beautiful target to observe or photograph.

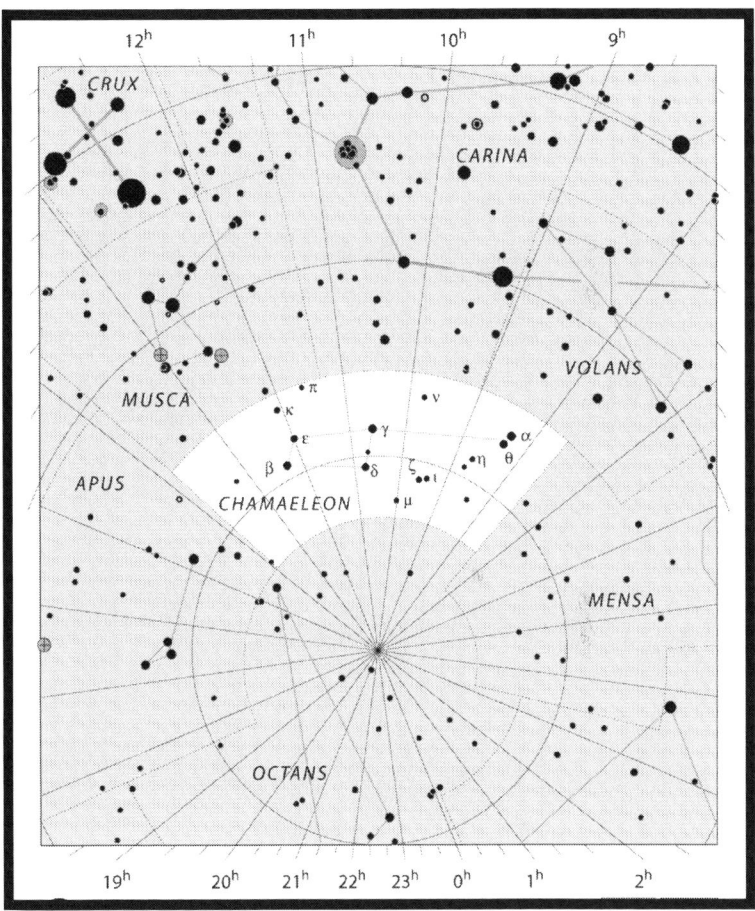

- Can be observed during **any** season of the year from the southern hemisphere
- There are **no** meteor showers associated with Chamaeleon
- The brightest star is **Alpha Chamaeleontis**

APUS

THE BIRD OF PARADISE

The Apus constellation is the 67th in size, and is located next to **Musca**. Just like the fly, Apus was named differently for many years (mostly Avis and/or Apis) until Nicolas-Louis de Lacaille, once again, decided to rename it.

Apus does not contain any galaxy or nebula worth observing, but does have two small globular clusters, **NGC 6101** and **IC 4499**.

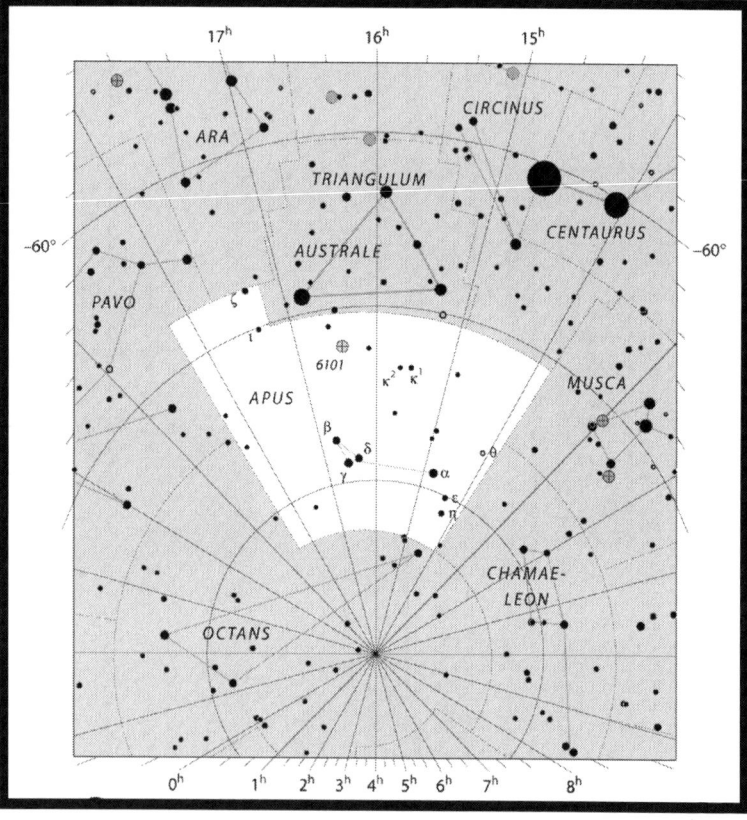

- Can be observed during **any** season of the year from the southern hemisphere
- There are **no** meteor showers associated with Apus
- The brightest star in the constellation is **Alpha Apodis**

THE SOUTHERN BIRDS

From **Musca**, this list moves in Westward direction around the south pole. Continuing in this fashion, we come across four constellations known as the Southern Birds: **Pavo**, **Tucana**, **Phoenix** and **Grus**.

Frederick de Houtman and Pieter Keyser most likely encountered these flying animals (except for the Phoenix) for the first time in their lives during their travels to the East Indies. Phoenix was simply named for the mystical animal, but is now part of the "Southern Birds" due to its location near the others.

The Southern Birds are some of the best constellations introduced by the Dutch, as they are brighter than most others and contain great deep sky activity.

Drawing of the Southern Birds

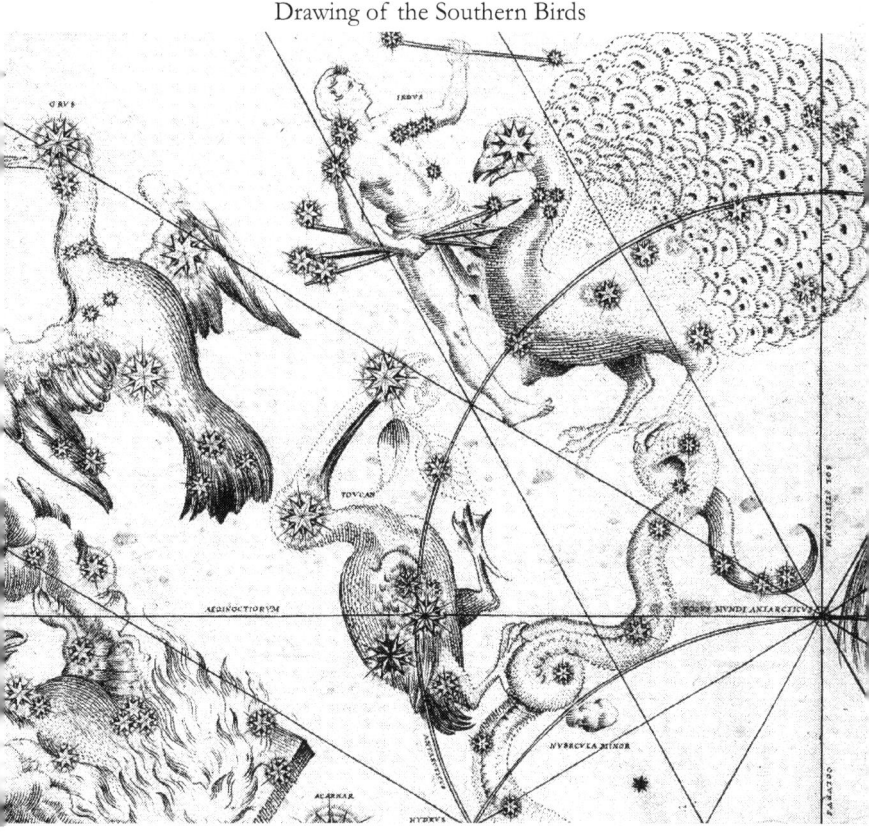

PAVO

THE PEACOCK

Pavo is thought to represent the green peacocks that Dutch explorers encountered several times during their travels through the East Indies.

Located in Pavo and with a magnitude of 5.4, **NGC 6752** is the third brightest globular cluster in the entire night sky. Several impressive galaxies can be found in the constellation of the peacock, like the two interacting galaxies **NGC 6872** and **IC 4970**, the spiral galaxy **NGC 6744**, and the triplet of interacting galaxies **IC 4686**, **IC 4687**, and **IC 4689**.

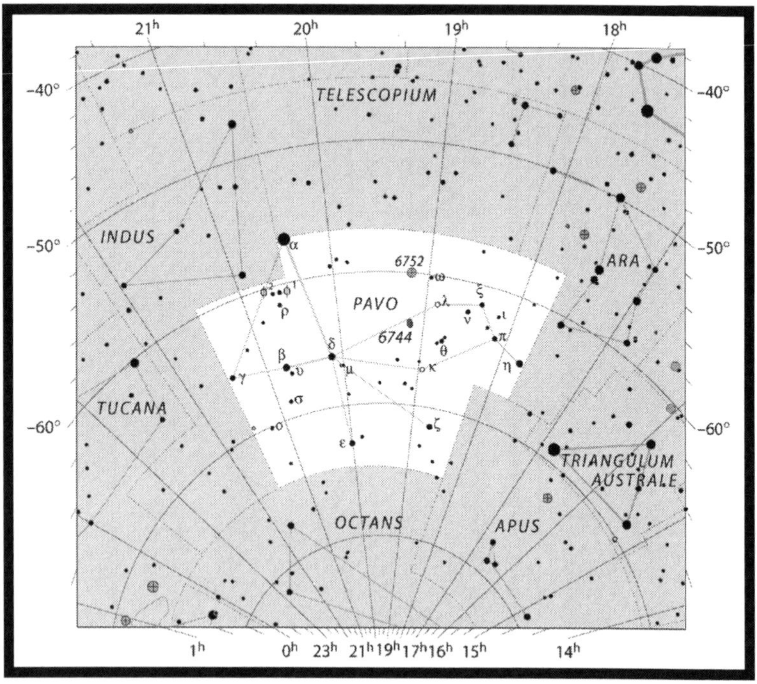

- Can be observed during **any** season of the year from the southern hemisphere
- The **Delta Pavonids** meteor shower happen in Pavo
- The brightest star in the constellation is **Alpha Pavonis**

TUCANA

THE TOUCAN

Towards the **Small Magellanic Cloud** is the constellation of the toucan bird: Tucana.
Besides the Small Magellanic Cloud, Tucana is home to the **Tucana Dwarf** galaxy, the spiral galaxy **NGC 406**, and a few other nebulous objects. It contains several star clusters as well, the most famous being **47 Tucanae**, the second brightest cluster in the entire sky after Omega Centauri.

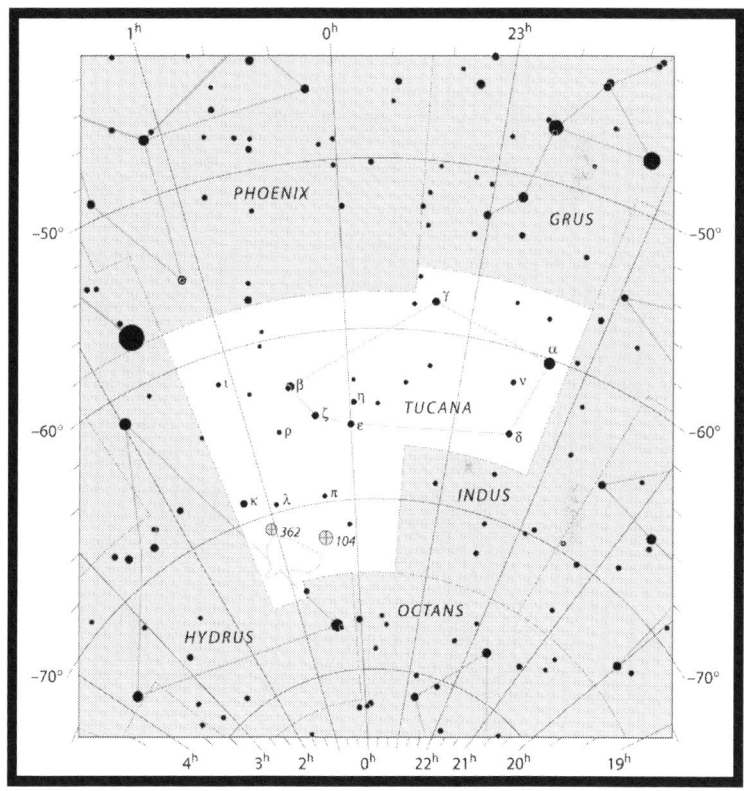

- Can be observed during **any** season of the year from the southern hemisphere
- There are **no** meteor showers associated with Tucana
- The brightest star in the constellation is **Alpha Tucanae**

PHOENIX

THE PHOENIX

Although a mythical bird, Phoenix is part of the Southern Birds constellations and is easily visible in the night sky. The Dutch may have wanted to depict a colorful eagle, but with **Aquila** already existing, it was replaced by a Phoenix. One of the best known features about Phoenix is that it is home to one of the most massive cluster of galaxies ever discovered: The **Phoenix Cluster**. In the center of the cluster is a fast growing supermassive black hole. **Robert's Quartet** is part of the constellation as well, and you can spot four galaxies, NGC 87, NGC 88, NGC 89 and NGC 92 close to one another.

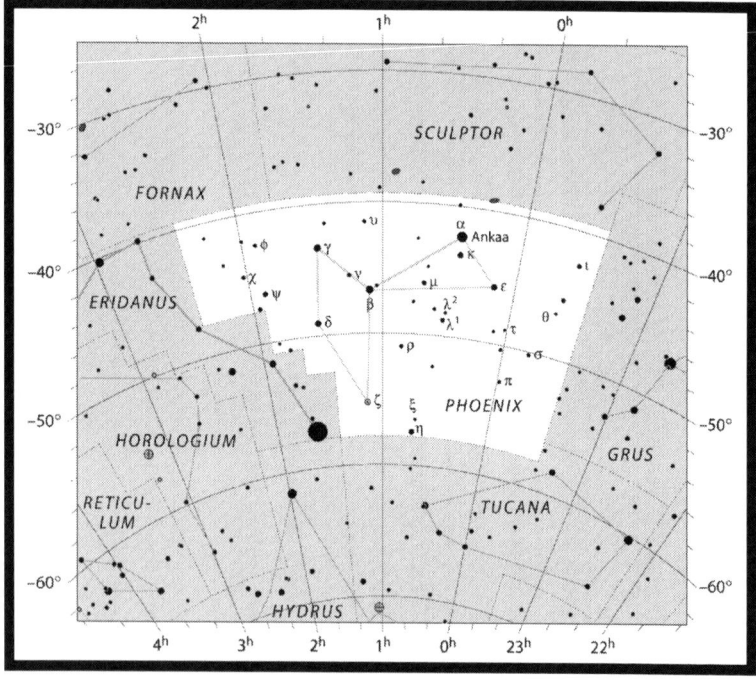

- Best observed in **Fall**
- The **Phoenicids** meteor shower peaks in early December
- The brightest star in the constellation is **Ankaa**

GRUS

THE CRANE

The last Southern bird is Grus, the crane. It is located just next to **Pisces Austrinus**, and one could imagine the crane trying to catch the fish as a reminder to finding the constellation in the night sky.

Grus is a medium size constellation and contains many deep sky objects, almost all of them are galaxies. Sadly, except for one or two, they are too faint to be properly observed through an eyepiece. The best ones are **NGC 7424**, **NGC 7213**, and **Grus Quartet**.

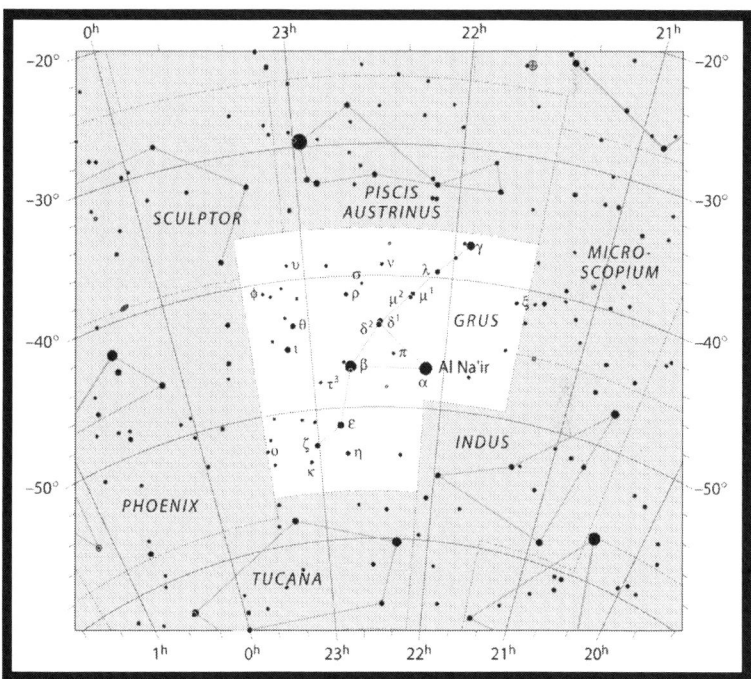

- Best observed in **Fall**
- There are **no** meteor showers associated with Grus
- The brightest star in the constellation is **Alnair**

VOLANS

THE FLYING FISH

On the other side of the south pole lies a small constellation called Volans, the flying fish.
It is home to several deep sky objects, but none are popular due to their difficulty of being seen. Volans contains two sets of double stars, Gamma Volantis and Epsilon Volantis, they offer a beautiful sight when seen through an eyepiece.

Too faint for visual observing but impressive to photograph is the Lindsay-Shapley Ring, an unbarred lenticular galaxy.

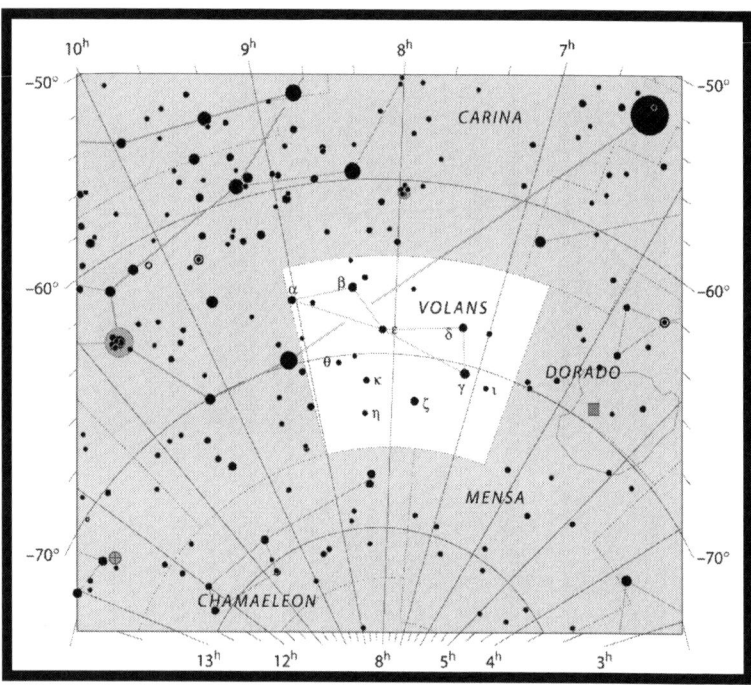

- Can be observed during **any** season of the year from the southern hemisphere
- There are **no** meteor showers associated with Volans
- The brightest star in the constellation is **Beta Volantis**

DORADO & HYDRUS

THE DOLPHINFISH AND THE MALE WATER SNAKE

Next to **Volans** is the constellation of Dorado, followed by Hydrus.
Dorado represents another water animal, the dolphinfish. The constellation is known for hosting most of the fourth largest galaxy in our galaxy group, the **Large Magellanic Cloud**. Although there are a couple of clusters, galaxies and nebulae in Dorado, its most interesting objects are within the Large Magellanic Cloud.
The **Tarantula** nebula (NGC 2070) is the largest one discovered near our galaxy, while the **Ghost Head** nebula (NGC 2080) is best known for its two white patches of gas that look like the eyes of a ghost.

Next to the dolphinfish is Hydrus, the "lesser" water snake. Hydrus lies between the Small and Large Magellanic Clouds, and is pretty void of bright deep sky objects. It was nicknamed "The male water snake" by Nicolas-Louis de Lacaille to differentiate it from the north water snake **Hydra**.
There are no meteor showers in either constellation.

Dorado (left) and Hydrus (right)

CAMELOPARDALIS

THE GIRAFFE

Introduced after the Dutch's voyage, the name Camelopardalis stands for camel and leopard, but the constellation represents what is now known as the giraffe. It is the 18th largest constellation in the night sky, but it is also one of the faintest, with only four stars brighter than magnitude 5. Camelopardalis contains few deep sky objects, mostly galaxies, such as the the intermediate spirals **NGC 2403** and **IC 342**, and the dwarf irregular **NGC 1569**.

An interesting fact is that the famous **Voyager 1** probe, launched on September 5, 1977, is currently heading in the direction of the giraffe.

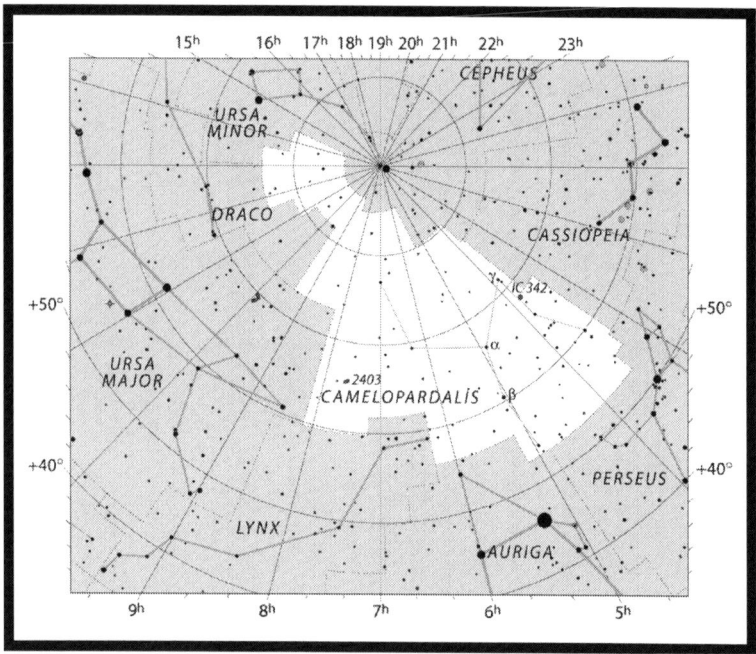

- Can be observed during **any** season of the year from the northern hemisphere
- The **October Camelopardalids** occur in Camelopardalis
- The brightest star is **Beta Camelopardalis**

MONOCEROS

THE UNICORN

Phoenix wasn't the only fantastic animal in Plancius' list of new constellations, another one, Monoceros, represents the Unicorn! Keyser and de Houtman did not encounter any in the East Indies, as this constellation, like **Camelopardalis**, was introduced after their voyage and lies in the Northern skies. Monoceros is a medium sized, faint constellation visible between **Orion**'s two dogs (**Canis Major** and **Canis Minor**).
There are many deep sky objects in the Unicorn, such as the **Rosette** nebula, the **Christmas Tree** cluster (and **Cone** nebula), the **Butterfly** nebula, and a Messier cluster, **M50**!

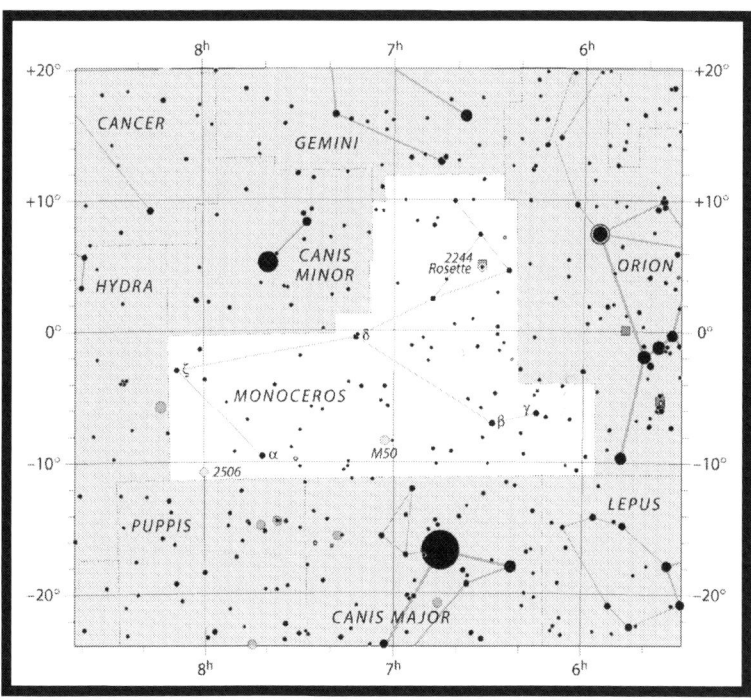

- Can be observed during **any** season of the year from the southern hemisphere
- There are **no** meteor showers associated with Musca
- The brightest star in the constellation is **Alpha Muscae**

These two constellations were introduced before de Houtman and Keyser's voyage to the East Indies. **Triangulum Australe** represents the level tool, while **Crux** is the southern cross, often associated with the cruxifixction of Jesus Christ.

COLUMBA

Columba was also drawn before the voyage, and is often linked to Noah's dove bringing the olive branch to the ark.

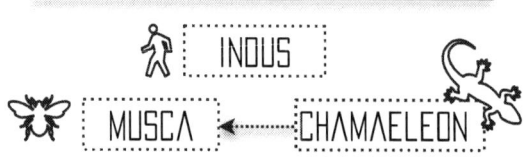

Indus is a native encountered in the Indies. **Chamaeleon**, the lizard, can be seen grabbing **Musca**, the fly, with its sticky tongue

APUS

Apus is the bird of paradise, the first bird the Dutch placed into the sky.

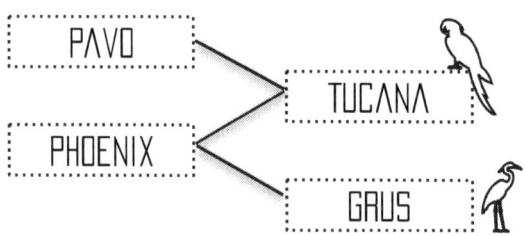

Pavo (the peacock), **Tucana** (the toucan), **Phoenix** (the phoenix), and **Grus** (the crane) make up the set of constellations called the "Southern Birds".

VOLANS ──── **DORADO** ──── **HYDRUS**

Volans (the flying fish), **Dorado** (the dolphinfish) and **Hydrus** (the male water snake) are three water-related constellations from the Dutch. They are close to each other.

CAMELOPARDALIS ──── **MONOCEROS**

Camelopardalis and **Monoceros** were both introduced by Plancius later, and are not related to the observations of the Dutch explorers, Frederick de Houtman and Pieter Keyser. Camelopardalis represents a type of camel/leopard, but was meant to represent a giraffe.
Monoceros is associated with a Unicorn.

We are now done with our fourth and last main astronomer to be credited for the introduction of the 88 constellations we know today.

The best way to learn Petrus Plancius' constellations is to remember which ones were introduced without the two explorers (the first three and the last two), and to then think about the natives they met when they arrived in the East Indies, followed by the 10 exotic animals they encountered during their travel.

For additional help in learning the constellations, we have included tables in the following pages, which you can use to add your own personal notes! That way, you can open up the end of this book at any time, take a glance at the constellations you are trying to remember, and see your own notes on how to memorize them.

INDEX - PTOLEMY

PAGE	CONSTELLATION	ASSOCIATED WITH	PERSONAL NOTES
14	ANDROMEDA	THE CHAINED LADY	
18	CEPHEUS	ANDROMEDA'S FATHER	
19	CASSIOPEIA	ANDROMEDA'S MOTHER	
20	PERSEUS	THE MONSTER SLAYER	
21	PEGASUS	THE FLYING HORSE	
22	CETUS	THE SEA MONSTER	
26	CAPRICORNUS	THE GOAT	
27	PISCES	THE TWO FISH	
28	AQUARIUS	THE WATER BEARER	
28	PISCES AUSTRINUS	THE FISH	
29	AQUILA	THE EAGLE	
30	SAGITTA	THE ARROW	

INDEX - PTOLEMY

PAGE	CONSTELLATION	ASSOCIATED WITH	PERSONAL NOTES
31	LYRA	THE LYRE	
32	DELPHINUS	THE DOLPHIN	
33	ERIDANUS	THE RIVER	
34	CYGNUS	THE SWAN	
38	ORION	THE HUNTER	
40	CANIS MAJOR	THE DOG	
40	LEPUS	THE HARE	
41	CANIS MINOR	THE LESSER DOG	
42	TAURUS	THE BULL	
44	AURIGA	THE CHARIOTEER	
45	GEMINI	THE TWINS	
46	SCORPIUS	THE SCORPION	

INDEX - PTOLEMY

PAGE	CONSTELLATION	ASSOCIATED WITH	PERSONAL NOTES
48	SAGITTARIUS	THE ARCHER	
50	ARIES	THE RAM	
51	TRIANGULUM	THE TRIANGLE	
52	OPHIUCHUS	THE SERPENT-BEARER	
52	SERPENS	THE SERPENT	
53	CORONA BOREALIS	THE NORTHERN CROWN	
54	CORONA AUSTRALIS	THE SOUTHERN CROWN	
55	VIRGO	THE GODDESS OF JUSTICE	
55	LIBRA	THE SCALES	
58	CORVUS	THE CROW	
59	CRATER	THE CUP	
60	HERCULES	HERAKLES	

INDEX - PTOLEMY

PAGE	CONSTELLATION	ASSOCIATED WITH	PERSONAL NOTES
62	CANCER	THE CRAB	
63	HYDRA	THE WATER SNAKE	
66	CENTAURUS	THE CENTAUR	
67	LUPUS	THE WOLF	
67	ARA	THE ALTAR	
68	EQUULEUS	THE SMALL HORSE	
69	LEO	THE LION	
70	COMA BERENICES	THE LION'S TUFT	
71	DRACO	THE DRAGON	
74	URSA MAJOR	THE GREATER BEAR	
75	URSA MINOR	THE LESSER BEAR	
76	BOÖTES	THE PLOWMAN	

INDEX - HEVELIUS

PAGE	CONSTELLATION	ASSOCIATED WITH	PERSONAL NOTES
84	CANES VENATICI	THE HUNTING DOGS	
86	LEO MINOR	THE LESSER LION	
86	LYNX	THE LYNX	
87	LACERTA	THE LIZARD	
88	VULPECULA	THE FOX	
89	SEXTANS	THE SEXTANT	
90	SCUTUM	SOBIESKI'S SHIELD	

INDEX - LACAILLE

PAGE	CONSTELLATION	ASSOCIATED WITH	PERSONAL NOTES
96	OCTANS	THE OCTANT	
97	MENSA	TABLE MOUNTAIN	
98	HOROLOGIUM	THE PENDULUM CLOCK	
98	RETICULUM	THE RETICULE	
99	TELESCOPIUM	THE TELESCOPE	

INDEX - LACAILLE

PAGE	CONSTELLATION	ASSOCIATED WITH	PERSONAL NOTES
99	MICROSCOPIUM	THE MICROSCOPE	
100	SCULPTOR	THE SCULPTOR'S STUDIO	
100	CAELUM	THE SCULPTOR'S CHISEL	
101	PICTOR	THE PAINTER'S EASEL	
102	FORNAX	THE FURNACE	
103	ANTLIA	THE AIR PUMP	
104	NORMA	THE SQUARE SET	
104	CIRCINUS	THE DRAWING COMPASS	
105	PYXIS	THE MARINER'S COMPASS	
108	CARINA	THE KEEL	
110	PUPPIS	THE POOP DECK	
111	VELA	THE SAILS	

INDEX - PLANCIUS

PAGE	CONSTELLATION	ASSOCIATED WITH	PERSONAL NOTES
116	TRIANGULUM AUSTRALE	THE LEVEL	
117	CRUX	THE SOUTHERN CROSS	
118	COLUMBA	NOAH'S DOVE	
119	INDUS	THE NATIVE	
120	MUSCA	THE FLY	
121	CHAMAELEON	THE CHAMELEON	
122	APUS	THE BIRD OF PARADISE	
124	PAVO	THE PEACOCK	

INDEX - PLANCIUS

PAGE	CONSTELLATION	ASSOCIATED WITH	PERSONAL NOTES
125	TUCANA	THE TOUCAN	
126	PHOENIX	THE PHOENIX	
127	GRUS	THE CRANE	
128	VOLANS	THE FLYING FISH	
129	DORADO	THE DOLPHINFISH	
129	HYDRUS	THE SOUTH WATER SNAKE	
130	CAMELOPARDALIS	THE GIRAFFE	
131	MONOCEROS	THE UNICORN	

CONCLUSION & OTHER WORK

We hope this guide was helpful to you, and will continue to be useful whenever you need a quick reminder on what's above.

If you enjoyed reading this book and would like to learn more about the hobby of astronomy, take a look at our other works!

With more than 200 pages dedicated to logging your stargazing sessions, **The Visual Astronomer's Journal** (right) is a portable notebook made for the purpose of recording observations, sketching your findings, and writing down wonderful memories created by this hobby.

Similarly to The Visual Astronomer's Journal, **The Astrophotographer's Journal** (left) is a notebook for people focusing on astrophotography. Instead of sketching your observations, you take photos of the night sky, then attach your work, jot down camera settings, and tips for the future.

Lastly, **The Astrophotographer's Guidebook** (center) is the perfect companion to any astrophotographer. It contains the best, and most impressive 60 targets to capture during the year, listed by season and difficulty, with our personal tips and tricks on getting the best final image possible!

To follow our adventures in the Nevada desert under thousands of stars, see our numerous tutorials, or a complete gallery of our images in high definition, visit our website **galactic-hunter.com** or watch our videos online at **YouTube.com/galactichunter**.

Clear Skies,
Galactic Hunter

CREDITS

Antoine & Dalia Grelin would like to thank all the following people whose images were illustrated in this book.

All constellation maps used in this book were created by IAU and Sky & Telescope magazine (Roger Sinnott & Rick Fiendberg).

Example #1: 2/3: Pages #2 and #3, full size image.
Example #2: 114-3: Page #114, third image.

Book Cover art:
Anieloeczek-stock - Old Book Cover I
Richard Rouse Bloxam - Urania's Mirror (1824)
Johannes Hevelius - Prodromus Astronomia - Volume III (1690)
2/3: Johannes Hevelius - Prodromus Astronomia - Vol III (1690)
8: Gregor Reisch - Ptolemy guided by the muse of Astronomy Urania (1508)
9: Johannes Hevelius - Philosophical Transactions (1670)
10: Nicolas-Louis de Lacaille - Coelum Australe Stelliferum (1763)
11: Johann Bayer - Uranometria (1603)
12-1: Johannes Honter - Imagines Constellationum Borealium (1541)
12-2: Johannes Hevelius - Prodromus Astronomia - Volume III (1690)
13: Andre Thevet Angoumoysin - Claude Ptolemee (1584)
17: Giuseppe Cesari - Perseus and Andromeda (1602)
24: Eustache Le Sueur - The Abduction of Ganymede (1650)
28: Alexander Jamieson - A Celestial Atlas (1822)
34: Johann Liss - The Fall of Phaethon (1624)
37: Nicolas Poussin - Blind Orion Searching for the Rising Sun (1658)
40: Richard Rouse Bloxam - Urania's Mirror (1824)
42: Lascaux Caves - Taurus (10,000+ BC)

43: Elihu Vedder - The Pleiades (1885).
46: Richard Rouse Bloxam - Urania's Mirror (1824)
52: Richard Rouse Bloxam - Urania's Mirror (1824)
55: Alexander Jamieson - A Celestial Atlas (1822)
57: Antonio del Pollaiuolo - Hercules and the Hydra (1475)
64: Pieter Paul Rubens - Hercules and the Nemean lion (1801)
67: Johannes Hevelius - Prodromus Astronomia - Vol III (1690)
73: Jean-François Millet - Arcas and Callisto (19th century)
82: Johannes Hevelius - Prodromus Astronomia - Volume III (1690)
83: Daniel Schultz - Johannes Hevelius (1658)
84: Richard Rouse Bloxam - Urania's Mirror (1824)
86: Alexander Jamieson - A Celestial Atlas (1822)
91-1: Johannes Hevelius - Uranographia (1687)
91-2: Unknown - The Battle of Vienna (Unknown)
94: Nicolas-Louis de Lacaille - Planisphère des Étoiles Australes (1756)
95: Melle Le Jeuneux - Abbé Nicolas-Louis de Lacaille (1762)
96: Cassini de Thury - Latitude Measurement (1740)
98: Nicolas-Louis de Lacaille - Planisphère des Étoiles Australes (1756)
99: Nicolas-Louis de Lacaille - Planisphère des Étoiles Australes (1756)
104: Nicolas-Louis de Lacaille - Planisphère des Étoiles Australes (1756)
105: Richard Rouse Bloxam - Urania's Mirror (1824)
107: Johannes Hevelius - Prodromus Astronomia - Volume III (1690)
108: NASA/ESA/Hubble - Mystic Mountains (2010)
109: Richard Rouse Bloxam - Urania's Mirror (1824)
114-1: Johann Bayer - Uranometria (1603)
114-2: Jacob Houbraken - Pieter Dirkszoon Hasselaer (1762)
114-3: David de Meyne - Frederick de Houtman (1617)
115: Unknown - Petrus Plancius (1622)
116: Johann Bode - Uranographia (1801)
119: Johann Bode - Uranographia (1801)
123: Johann Bode - Uranographia (1801)

Printed in Great Britain
by Amazon